Sir Edward Hertslet

The Map of Africa by Treaty

Vol. III

Sir Edward Hertslet

The Map of Africa by Treaty
Vol. III

ISBN/EAN: 9783744751834

Printed in Europe, USA, Canada, Australia, Japan

Cover: Foto ©berggeist007 / pixelio.de

More available books at **www.hansebooks.com**

THE
MAP OF AFRICA BY TREATY.

(SECOND AND REVISED EDITION.)

VOL. III.
APPENDIX,
ALPHABETICAL INDEX,
AND
CHRONOLOGICAL LIST.

With Two Maps.

BY

SIR EDWARD HERTSLET, K.C.B.

(*Late Librarian and Keeper of the Papers, Foreign Office.*)

LONDON:
PRINTED FOR HER MAJESTY'S STATIONERY OFFICE,
BY HARRISON AND SONS, ST. MARTIN'S LANE,
PRINTERS IN ORDINARY TO HER MAJESTY.

And to be purchased, either directly or through any Bookseller, from
EYRE & SPOTTISWOODE, EAST HARDING STREET, FLEET STREET, E.C.; or
JOHN MENZIES & Co., 12, HANOVER STREET, EDINBURGH, and
90, WEST NILE STREET, GLASGOW; or
HODGES, FIGGIS, & Co., Limited, 104, GRAFTON STREET, DUBLIN.

1896.
Price (with Vols. I and II) Thirty-one Shillings and Sixpence.

(Wt. 21059 500 8|96—H & S 3884.)

APPENDIX.

App.						Page
1	1885.	1 June.	Treaty..	..	National African Company and Sokoto (1). Jurisdiction over Foreigners, &c.	972
2	,,	13 ,,	Treaty..	..	National African Company and Gandu (1). Transfer of rights, &c.	974
3	1888.	2/9 Feb.	Agreement	..	British and Foreign Governments. Gulf of Tajoura and Somali Coast	976
4	1890.	20 Jan.	Treaty..	..	Royal Niger Company and Boussa (Borgu). British Protection, &c.	981
5	,,	7 Apr.	Treaty..	..	Royal Niger Company and Gandu (2). Protection, &c.	983
6	,,	15 ,,	Treaty..	..	Royal Niger Company and Sokoto (2). Jurisdiction, &c. ..	984
7	,,	17 Nov.	Exchange of Notes		France and Germany. Madagascar, Zanzibar, and Mafia	985
8	1891.	Feb.	Conditions	..	Extension of Operations of British South Africa Company to the North of the Zambesi.. ..	987
9	,,	27 June.	Proclamation..		Cape of Good Hope. British Jurisdiction within protected Territories. (Bechuanaland, Tati District, &c.)	990
10	1892.	8 Feb.	Notification ..		Free Port of Zanzibar	992
11	,,	22 June.	Notification ..		British Protectorate of Zanzibar placed under Free Trade provisions of "Berlin Act" ..	993
12	1893.	29 May.	Provisional Agreement		Mwanga, King of Uganda, and Sir Gerald Portal. British Protection, &c...	995
13	1894.	5 Jan.	Declaration	..	France. Submission of Chiefs, &c., of Dahomey. Country on left bank of the Ouemé under French Protection ˙ .. Note	998
13	,,	29 Jan.	Treaty..	..	France and Abomey. French Protectorate .. .. Note	998
13	,,	4 Feb.	Declaration	..	France. French Protectorate over Allada .. .. Note	998
13	,,	22 June.	Decree..	..	France. Coast Territory of Dahomey a French Colony. Note	998
14	,,	4 Feb. / 15 Mar.	Protocol	..	France and Germany. Boundaries. Cameroons. French Congo. Lake Tchad	999

LIST OF ADDITIONAL DOCUMENTS.

[Arranged in Order of Date.]

App.						Page
15	1894.	24 Mar.	Declaration	..	Congo and Portugal. Boundary. Lunda Region	1004
16	,,	12 May.	Agreement	..	Great Britain and Congo Free State. Spheres of Influence in East and Central Africa ..	1008
17	,,	18 June.	Notification	..	British Protectorate over Uganda	1016
18	,,	22 ,,	Declaration	..	Great Britain and Congo Free State. Withdrawal of Art. III of Agreement of 12 May, 1894	1017
19	,,	4 July.	Treaty ..	..	Royal Niger Company and Gandu (3). Jurisdiction over Foreigners, &c.	1018
20	,,	18 ,,	Order in Council		Matabeleland. Boundaries, &c.	1020
21	,,	14 Aug.	Agreement	..	France and Congo Free State. Boundaries	1021
22	,,	27 Aug.	Treaty ..	..	Great Britain and Uganda. British Protectorate	1023
23	,,	— Sept.	Agreement	..	Germany and Portugal. Spheres of Influence. East Africa. Kionga, &c.	1024
24	,,	24 Nov.	Agreement	..	British South Africa Company. British Central Africa North of the Zambesi	1025
25	,,	10 Dec.	Convention	..	Great Britain and South African Republic. Swaziland	1029
26	1895.	7 Jan.	Declaration	..	Great Britain and Portugal. Reference to Arbitration of Boundary Dispute. Manica plateau .	1037
27	,,	9 ,,	Treaty*		Congo and Belgium. Cession ..	1041
28	,,	11 ,,	Declaration*	..	Belgium. Neutrality. Congo State	1043
29	,,	21 ,,	Agreement	..	Great Britain and France. Boundary, North and East of Sierra Leone	1048
30	,,	5 Feb.	Arrangement*		Belgium and France. French right of pre-emption over territories of Congo State	1059
31	,,	5 ,,	Declaration*	..	Belgium and France. Possessions in Stanley Pool	1061
32	,,	24 ,,	Supplementary Convention		Spain and Morocco. Indemnities, &c.	1062
33	,,	13 March.	Agreement	..	Great Britain and Morocco. Purchase of Property of N. W. Africa Company in Terfaya (Cape Juby)	1064
34	,,	23 April.	Proclamation ..		Natal. British Sovereignty over Territories of certain Native Chiefs in Zululand	1067
35	,,	11 June.	Notification	..	British Protectorate. Amatongaland	1068

* Treaty 9 Jan., 1895, not yet ratified (1 Feb., 1896).

LIST OF ADDITIONAL DOCUMENTS.

[Arranged in Order of Date.]

App.						Page
36	1893.	15 June.	Notification	..	British Protectorate. British East Africa	1069
37	„	25 „	Agreement	..	Italy and Egypt. Boundary between the Baraka and the Red Sea	1072
38	„	Sept.–Oct.	Exchange of Notes		Great Britain and Portugal. Boundary. Tongaland and Portuguese possessions ..	1075
39	„	4 Nov.	Notification	..	Great Britain. Non-recognition of certain concessions made to Amatongaland Exploration Co.	1078

[Sokoto.]

App. 1.—*TREATY. National African Company and Sokoto. Transfer of Rights, &c.* 1st *June*, 1885.*

COPY of English duplicate of Treaty between Umoru, King of the Mussulmans of the Soudan and Sultan of Sokoto, for himself and Chiefs, on the one part, and those Europeans trading on the Kworra and Benué, under the name of the "National African Company (Limited)," on the other part.

ART. I. For the mutual advantage of ourselves and people, and those Europeans trading under the name of the "National African Company (Limited)," I, Umoru, King of the Mussulmans of the Soudan, with the consent and advice of my Council, grant and transfer to the above people, or other with whom they may arrange, my entire rights to the country on both sides of the River Benué and rivers flowing into it throughout my dominions for such distance from its and their banks as they may desire.

ART. II. We further grant to the above-mentioned Company the sole right, among foreigners, to trade in our territories, and the sole right, also among foreigners, to possess or work places from which are extracted articles such as lead and antimony.

ART. III. We further declare that no communication will be held with foreigners coming from the rivers except through the above-mentioned Company.

ART. IV. These grants we make for ourselves, our heirs, and successors for ever, and declare them to be irrevocable.

ART. V. The Europeans above named, the National African Company (Limited), agree to make Umoru, Sultan of Sokoto, a yearly present of goods to the value of 3,000 bags of cowries, in return for the above grants.

Signed and sealed at Wurnu, the 1st June, 1885.

(Signature of the Sultan in Arabic.)
(Great seal of the Empire of Sokoto.)

* Confirmed by Treaty, 15th April, 1890. Page 984.

[Sokoto.]

For the National African Company (Limited).
                    JOSEPH THOMSON, *F.R.G.S.*
Witnesses:
    W. J. Seago.
    D. Z. Viera.
    T. Joseph.

[An Arabic duplicate was at the same time executed by both parties.]

App. 2.—*TREATY. National African Company and Gandu. Transfer of Rights, &c.* 13*th June*, 1885.*

ENGLISH Duplicate of Treaty between Maliké, King of Gandu, for himself and Chiefs, on the one part, and those Europeans trading on the Kworra and Benué, under the name of the National African Company (Limited), on the other part.

ART. I. For the mutual advantage of myself and people, and of those Europeans trading under the name of "the National African Company (Limited)," I, Maliké, of Gandu, with the consent and advice of my Council, grant and transfer to the above people, or others with whom they may arrange in future, my entire rights, absolutely, to the country on both sides of the Rivers Benué and Kworra, for a distance of ten hours' journey inland, or such other distance as they may desire, from each bank of both rivers throughout my dominions.

ART. II. I further grant to the above-mentioned people the sole right, among foreigners, to trade in my territories, and the sole right, also among foreigners, to possess or work places from which are taken articles such as lead and antimony.

ART. III. I further declare that no communication will be held with foreigners coming from the rivers except through the above-mentioned people.

ART. IV. These grants I make for myself, my heirs and successors, and declare them to be unchangeable and irrevocable.

ART. V. The Europeans above mentioned, under the name of the National African Company (Limited), agree to make a yearly present for the above grants to Maliké, King of Gandu, of goods to the value of 2,000 bags of cowries.

* Confirmed by Treaty 7 April, 1890. Page 983.

[Gandu.]

Signed at Gandu on the 13th June, 1885.
For the National African Company (Limited).

JOSEPH THOMSON, *F.R.G.S.*

[Duly executed in Arabic by the Government of Gandu.]
Witnesses:
W. J. SEAGO.
D. Z. VIERA.
T. JOSEPH.

[An Arabic duplicate was at the same time executed by both parties.]

App. 3.—AGREEMENT *between the British and French Governments with regard to the Gulf of Tajourra and the Somali Coast,* $\frac{2nd}{9th}$ *February,* 1888.*

(1.) *M. Waddington to the Marquis of Salisbury.*

(Translation.)

London, February 2, 1888.

M. le Marquis,

The Government of the French Republic and the Government of Her Britannic Majesty being desirous of arriving at an agreement with regard to their respective rights in the Gulf of Tajourra, and on the Somali Coast, I have had the honour to address your Lordship on this subject on several occasions. After a friendly interchange of views we yesterday agreed on the following arrangement:—

1. The Protectorates exercised, or to be exercised, by France and Great Britain shall be separated by a straight line starting from a point on the coast situated opposite the wells of Hadou, and leading through the said wells to Abassouen; from Abassouen the line shall follow the caravan road as far as Bia-Kabouba, and from this latter point it shall follow the caravan route from Zeyla to Harrar, passing by Gildessa. It is expressly agreed that the use of the wells of Hadou shall be common to both parties.

2. The Government of Her Britannic Majesty recognises the Protectorate of France over the coasts of the Gulf of Tajourra, including the group of the Mushah Islands and the Islet of Bab, situated in the gulf, as well as over the inhabitants, tribes, and fractions of tribes situated to the west of the line above mentioned.

The Government of the French Republic recognise the Protectorate of Great Britain over the coast to the east of the above line as far as Bender Ziadeh, as well as over the inhabitants, tribes, and fractions of tribes situated to the east of the same line.

* Parliamentary Paper, "France No. 1 (1894)."

App. 3] GREAT BRITAIN AND FRANCE. [2/9 Feb., 1888.
[Gulf of Tajourra. Somali Coast.]

3. The two Governments engage to abstain from any action or intervention, the Government of the Republic to the east of the above line, the Government of Her Britannic Majesty to the west of the same line.

4. The two Governments engage not to endeavour to annex Harrar, nor to place it under their Protectorate. In taking this engagement the two Governments do not renounce the right of opposing attempts on the part of any other Power to acquire or assert any rights over Harrar.

5. It is expressly understood that the caravan route from Zeyla to Harrar, by way of Gildessa, shall remain open throughout its extent to the commerce of the two nations as well as to that of the natives.

6. The two Governments engage to take all necessary measures to prevent the slave trade, and the importation of gunpowder and arms in the territories subject to their authority.

7. The Government of Her Britannic Majesty engage to treat with consideration ("bienveillance") those persons, whether chiefs or members of the tribes placed under their Protectorate, who had previously adopted the French Protectorate. The Government of the Republic, on their part, take the same engagement with regard to the persons and tribes placed henceforth under their Protectorate.

I shall be grateful if, in acknowledging the receipt of this note, your Lordship will record officially the Agreement which we have concluded in the names of our respective Governments.

Accept, &c.,
WADDINGTON.

(2.) *The Marquis of Salisbury to M. Waddington.*

Foreign Office, February 9, 1888.

M. l'Ambassadeur,

I have the honour to acknowledge the receipt of your Excellency's note of the 2nd instant, reciting the arrangement upon which we have agreed with regard to the respective rights of Great Britain and France in the Gulf of Tajourra and on the Somali Coast.

The provisions of this arrangement are as follows :—

1. The Protectorates exercised, or to be exercised, by Great Britain and France shall be separated by a straight line starting from a point on the coast opposite to the wells of Hadou and passing through the said wells to Abassouen; from Abassouen the line shall follow the caravan road as far as Bia-Kabouba, and from this latter point it shall follow the caravan road from Zeyla to Harrar, passing through Gildessa. It is expressly agreed that the use of the wells of Hadou shall be common to both parties.

2. Her Britannic Majesty's Government recognise the Protectorate of France over the coasts of the Gulf of Tajourra, including the Group of the Mushah Islands and the Islet of Bab, situated in the gulf, as well as over the inhabitants, tribes, and fractions of tribes situated to the west of the line above mentioned.

The Government of the French Republic recognise the Protectorate of Great Britain over the coast to the east of the above line as far as Bender Ziadeh, as well as over the inhabitants, tribes, and fractions of tribes situated to the east of the same line.

3. The two Governments pledge themselves to abstain from taking any action or exercising any intervention, the Government of the Republic to the east of the above line, Her Britannic Majesty's Government to the west of the same line.

4. The two Governments engage not to endeavour to annex Harrar, nor to place it under their Protectorate. In taking this engagement the two Governments do not renounce the right of opposing attempts by any other Power to acquire or assert any rights over Harrar.

5. It is expressly agreed that the caravan road from Zeyla to Harrar, by way of Gildessa, shall remain open in its entire extent to the commerce of the two nations, as well as to that of the natives.

6. The two Governments engage to take all necessary

App. 3] GREAT BRITAIN AND FRANCE. [$^2_9$ Feb., 1888.

[Gulf of Tajourra. Somali Coast.]

measures to prevent the slave trade and the importation of gunpowder and arms in the territories subject to their authority.

7. The Government of Her Britannic Majesty engages to treat with consideration ("bienveillance") those persons, whether chiefs or members of the tribes placed under their Protectorate, who had previously adopted the French Protectorate. The Government of the Republic, on their part, take the same engagement with regard to the persons and tribes henceforth placed under their Protectorate.

I have the honour to state that the arrangement recited in your Excellency's note, of which the above is a textual translation, is accepted by Her Majesty's Government, and will be considered by them as binding upon the two countries from the present date.

In doing so, I will add, for the sake of record, that I understand the third clause of the Agreement to preclude the granting by either party of protection to natives within the Protectorate of the other party; and that I gathered in conversation that your Excellency concurred with me in that opinion.

I have, &c.,

SALISBURY.

Reservation of Rights of the Sultan of Turkey.

(3.) *The Marquis of Salisbury to M. Waddington.*

Foreign Office, February 9, 1888.

M. l'Ambassadeur,

With reference to the note which I have this day addressed to your Excellency, accepting, on behalf of Her Majesty's Government, the arrangement agreed upon between us respecting the British and French Protectorates in the Gulf of Tajourra and on the Somali coast, I think it right to remind your Excellency that I received some months ago a request from the Turkish Ambassador at this Court that in any understanding which might be arrived at on this subject the rights of His Imperial Majesty the Sultan might be respected.

I assured his Excellency, in reply, that the British Government would carefully abstain in the future, as in the past, from any interference with the just rights of the Sultan, and that I was convinced that the Government of the French Republic would act in a similar spirit.

I have, &c.,
SALISBURY.

App. 4.—*TREATY. Royal Niger Company and Boussa (Borgu). British Protection.* 20th January, 1890.*

TREATY entered into between the Emir and Chiefs of Boussa (or Borgu) on behalf of themselves and their successors, for ever, and the Royal Niger Company (Chartered and Limited), hereinafter called "the Company," on behalf of themselves, their successors, and assigns.

WE, the Emir and Chiefs of Boussa (or Borgu), in Council assembled (representing our country, its dependencies, and tributaries on both banks of the River Niger, and as far back as our dominion extends, in accordance with our laws and customs), do hereby agree, on behalf of ourselves and of our successors, for ever:—

Firstly. To observe faithfully the Agreement entered into between us and the Company (then known as the National African Company, Limited), and dated the 12th day of November, 1885.

Secondly. To grant to the Company full and absolute jurisdiction over all foreigners to our territories—that is to say, over all persons within the territories who are not our native-born subjects. Such jurisdiction shall include the right of protection of such foreigners, of taxation of such foreigners, and of political, criminal, and civil jurisdiction over such foreigners.

Thirdly. That we will not at any time whatever cede any of our territories to any other person or State, or enter into any Agreement, Treaty, or arrangement with any foreign Government, except through and with the consent of the Company; or, if the Company should at any time so desire, with the consent of the Government of Her Majesty the Queen of Great Britain and Ireland, and Empress of India.

Fourthly. To place our territories, if and when called upon to do so by the Company, under the protection of the flag of Great Britain.

* On the 20th October, 1894, it was formally notified to the German Government that a British Protectorate had been established over Borgu.

[Boussa (Borgu).]

[Engagement of the Royal Niger Company.

1. To admit to the territories of Boussa (or Borgu) any foreigner who may desire to go there, subject to such necessary restrictions as may be necessary in the interests of peace and order.

2. To permit all such foreigners to trade freely, subject to the payment of such taxation as may be necessary for administrative purposes in Boussa (or Borgu), or for the general administration of the Company.

3. To do our utmost to promote the prosperity and wealth of Boussa (or Borgu), and to develop and open up that country, and to do the utmost in our power to promote peace, order, and good government, and the general progress of civilization.

4. To pay to the Emir of Boussa (or Borgu) a yearly sum of 50 bags, native value, in any class of goods, to be taken at the market value of the place where and when the payment is made.]

For the Royal Niger Company (Chartered and Limited),

                                    WILL LISTER.

Signatures of the Emir and Chiefs:

    DACHTRA (or DAGGA), *Emir of Boussa (or Borgu)*, his X mark.
    MOMO (eldest son of Emir), *ditto*.
    MUSA, *Eyusu, ditto*.
    SERIKIN RUA, *Chief, ditto*.

We, the undersigned, are witnesses to the above signatures and marks.

    GILA.
(Arabic signature.)

I, William Reffle, do hereby certify that the above has been faithfully interpreted to the Emir and Chiefs of Boussa (or Borgu), and understood by them in every sense.

                                      W. REFFLE.

Done in triplicate at Boussa, this 20th day of January, 1890.

App. 5.—*TREATY. Royal Niger Company and Gandu. Protection. Jurisdiction over Foreigners, &c. 7th April,* 1890.*

LITERAL translation of second Treaty, in Arabic, between Maliké, King of Gandu, for himself and Chiefs, on the one part, and the Royal Niger Company (Chartered and Limited) on the other part.

BE it known that I, Maliké, King of Gandu, am desirous of introducing European trade in all parts of my dominions, so as to increase the prosperity of my people, and knowing that this cannot be effected except by securing to foreigners the protection of European government, with power of exercising jurisdiction over foreigners as is the custom with them, also with power of levying taxes upon foreigners as may be necessary for the exercise and support of this jurisdiction: I, Maliké, King of Gandu, with the consent and advice of my Council, agree and grant to the Royal Niger Company (Chartered and Limited), formerly known as the "National African Company (Limited)," full and complete power and jurisdiction over all foreigners visiting and residing in any part of my dominions. I also grant you jurisdiction and full rights of protection over all foreigners, also power of raising taxes of any kind whatsoever from such foreigners.

No person shall exercise any jurisdiction over such foreigners nor levy any tax whatsoever on such foreigners than the Royal Niger Company (Chartered and Limited).

These grants I make for myself, my heirs, and successors, and declare them to be unchangeable and irrevocable for ever.

I further confirm the Treaty made by me with the National African Company (Limited), now known as the "Royal Niger Company (Chartered and Limited)," in the month of June, according to European reckoning, 1885 (page 974).

Dated at Gandu, this 7th day of April, 1890.

* See also Treaty, 4th July, 1894. Page 1018.

[Sokoto.]

App. 6.—*TREATY. Royal Niger Company and Sokoto. Jurisdiction over Foreigners, &c. 15th April, 1890.*

LITERAL translation of second Treaty, in Arabic, between Umoru, King of the Mussulmans of the Soudan, and Sultan of Sokoto, on the one part, and the Royal Niger Company (Chartered and Limited), on the other part.

BE it known that I, Umoru, King of the Mussulmans, am desirous of introducing European trade in all parts of my dominions, so as to increase the prosperity of my people, and knowing that this cannot be effected except by securing to foreigners the protection of European government, with power of exercising jurisdiction over foreigners, as is the custom with them; also with power of levying taxes upon foreigners as may be necessary for the exercise and support of this jurisdiction: I, Umoru, King of the Mussulmans of the Soudan, with the consent and advice of my Council, agree and grant to the Royal Niger Company (Chartered and Limited)—formerly known as the "National African Company (Limited)"—full and complete power and jurisdiction over all foreigners visiting or residing in any part of my dominions. I also grant you jurisdiction and full rights of protection over all foreigners; also power of raising taxes of any kind whatsoever from such foreigners.

No person shall exercise any jurisdiction over such foreigners, nor levy any tax whatsoever on such foreigners than the Royal Niger Company (Chartered and Limited).

These grants I make for myself, my heirs, and successors, and declare them to be unchangeable and irrevocable for ever.

I further confirm the Treaty made by me with the National African Company (Limited)—now known as the "Royal Niger Company (Chartered and Limited)"—in the month of June, according to European reckoning, 1885 (page 972).

Dated at Wurnu, this 15th day of April, 1890.

App. 7.—*EXCHANGE OF NOTES between the French and German Governments respecting the recognition of the French Protectorate over Zanzibar, and the German acquisition of the Continental Possessions of the Sultan of Zanzibar and of the Island of Mafia.* 17th *November,* 1890.

(1.)—*M. Herbette, French Ambassador at Berlin, to Baron de Marschall, German Minister for Foreign Affairs.*

(Translation.)

Berlin, 17th *November,* 1890.

In the course of certain discussions, which we had together in the month of August last, on the reciprocal relations of Germany and France, on the East Coast of Africa, your Excellency declared that the Imperial Government was disposed to recognize the Protectorate of France over Madagascar with all its consequences.

On my side, I was about to give you, at our interview on the 6th instant, the assurance that the French Government would raise no objection to the acquisition by Germany of the continental portion of the Dominions of the Sultan of Zanzibar, as well as of the Island of Mafia.

It is, moreover, agreed that German subjects in Madagascar, and French subjects in the territories ceded to Germany by the Sultan of Zanzibar, shall enjoy in every respect most-favoured-nation treatment.

With the view of establishing definitively the agreement between the two Governments on these two points, I have the honour to address to your Excellency the present communication, and I beg that you will cause the receipt thereof to be acknowledged to me.

HERBETTE, *Baron.*

(2.)—*Baron Marschall to M. Herbette.*

(Translation.)   Berlin, 17th *November,* 1890.

The undersigned has the honour to acknowledge the receipt

[Madagascar. Zanzibar. Mafia.]

to his Excellency the Ambassador Extraordinary and Minister Plenipotentiary of the French Republic, M. Jules Herbette, of the letter which he did him the honour to address to him this day and to make known to him that the Imperial Government adheres to the declarations therein contained. It results from it that the French Republic offer no objection to the acquisition by Germany of the Continental Possessions of the Sultan of Zanzibar and of the Island of Mafia, and that Germany, on her part, recognizes the Protectorate of France over Madagascar, with all its consequences.

It is, moreover, expressly agreed that German subjects in Madagascar, and French citizens in the above-mentioned territories, which the Sultan of Zanzibar cedes to Germany, shall enjoy most-favoured-nation treatment.

MARSCHALL.

[North of the Zambezi.]

App. 8.—*CONDITIONS on extending the Field of the Operations of the British South Africa Company to the North of the Zambezi. February,* 1891.*

Extension of field of Company's operations North of the Zambezi.

THE Charter of the British South Africa Company (**No. 37**) shall extend over the territory under British influence north of the Zambezi and south of the territories of the Congo Free State and the German sphere, and accordingly the Company is hereby granted powers necessary for the purposes of good government and the preservation of public order in and for the protection of the said territory under British influence, but subject to the following conditions :—

Nyasaland excluded from Field of Operations.

1. The said field of operations shall not include Nyasaland.

Definition of Nyasaland Territory.

The territory defined by that name will be bounded, where it adjoins the Chartered territory, by a frontier which, starting on the south from a point where the boundary between the British and Portuguese spheres is intersected by the boundary of the Conventional line of the Berlin Act (**No. 17**), will follow that line to the point where it meets the geographical line of the Congo Basin, and will thence follow the latter line to the point where it reaches the boundary between the British and German spheres.

Powers of Government and Administration.

2. As regards the powers of government and administration by the Company, the Secretary of State shall, pursuant to the power reserved to him by Article IV of the Charter (**No. 37**) subject them to the condition that, until the 1st January, 1894, or until such earlier date as he shall direct, they shall be exercised for the Company by Her Majesty's Commissioner

* Parl. Pap., Africa, No. 2 (1895). See also Memorandum, 24th November, 1894. Page 1025.

for Nyasaland in consultation with the Company, and accordingly, in this respect, the Company's officers shall be subordinate to the Commissioner.

After the 1st January, 1894, the arrangement shall be renewable, at the discretion of Her Majesty's Government, for a further period not exceeding two years.*

Preservation of Peace and Order. Police Force.

3. The duty of preserving peace and order incumbent on the Company under Article X of the Charter (**No. 37**) shall devolve on the said Commissioner so long as Article II hereof is in force. The Commissioner shall have the control of the police force, the establishment of which is authorized by Article X of the Charter, with power to employ it at his discretion in any part of the Company's field of operations north of the Zambezi and in Nyasaland.

Payment for Police Force, including armed Boats, not less than 10,000l. a year.

The Company shall raise, equip, and maintain (providing the necessary barrack accommodation) the police force (under which head armed boats shall be comprised), and defray all expenses connected with its employment, expending for these purposes through the said Commissioner not less than 10,000l. a year.

The said Commissioner shall be consulted as to the organization of the police, and especially as to the appointment by the Company of its officers.

Administration of Justice.

4. Justice to the peoples and inhabitants within the Company's field of operations north of the Zambezi, under Article XIV of the Charter (**No. 37**) shall be administered by the said Commissioner so long as Article II hereof is in force.

5. The administration of justice shall be in conformity with the Africa Order in Council of the 15th October, 1889,† under

* See Memorandum, 24th November, 1894. Page 1025.
† H. T., vol. xviii, p. 1.

which judicial powers will be conferred on the said Commissioner (so long as Article II hereof is in force), and on such other officers who may be employés of the Company as the Secretary of State shall, at the request of the Company nominate.

Goods passsing through Nyasaland.

6. Goods passing through Nyasaland to or from the Chartered territory shall be treated as goods in transit, and shall be free from duty.

If, for the sake of convenience, duties are levied on them on the Nyasaland frontiers, they shall be accounted for to the Company.

Payment of Expenses of Administration in the Chartered Territory.

7. All expenses connected with the administration in the Chartered territory shall be borne by the Company either by a fixed payment, or by liquidation of accounts rendered by the Commissioner, but no expense beyond the before-mentioned 10,000*l.*, except for travelling expenses of the Commissioner and his agents, shall be incurred without the previous sanction of the Company.

Nyasaland. Material of War and Steamers belonging to the African Lakes Company.

8. The Company shall make arrangements under which the said Commissioner shall, in Nyasaland, be authorized to make use of material of war belonging to the African Lakes Company in case of necessity, and under which he shall be empowered to use, free of charge, for administrative purposes, the steamers belonging to that Company on Lake Nyasa, with due precautions against unreasonable interference with their employment for the Company's trade.

Foreign Office, February 1891.

[This Agreement was renewed for two years from the 1st January, 1894. See Memorandum, 24th November, 1894, p. 1025.]

27 June, 1891.] GT. BRITAIN (BECHUANALAND). [App 9
[Tati District, &c.]

App. 9.—*PROCLAMATION of the Governor of the Cape of Good Hope, &c., with regard to the exercise of British Jurisdiction within certain protected Territories. Bechuanaland, Tati District, &c.* 27th June, 1891.

[After Preamble].

1. Now, therefore, I do hereby proclaim, declare, and make known that the Resident Commissioners, Assistant Commissioners, and Magistrates within the protected territories defined in Her Majesty's Order in Council, dated the 9th day of May, 1891,* shall exercise jurisdiction and authority as follows:—

(a.) The Resident Commissioner for Bechuanaland and the Tati District, within the Tati District, the territory known as the disputed territory lying between the Shashi and Macloutsie Rivers, excepting the area included in the Tuli District, and the territories lying between the Crown Colony of British Bechuanaland and the 22nd parallel of south latitude, and also such territories north of the 22nd degree as belong to the Chief Khama of the Bamangwato.

(b.) The Resident Commissioner for Mashonaland shall exercise jurisdiction within the territories north of the 22nd parallel of south latitude, excluding the disputed territory, and all territories belonging to the Chief Khama of the Bamangwato, as well as the Tati District, and for an area of 10 miles round Tuli Fort.

2. The Assistant Commissioner and Magistrate for Bechuanaland shall exercise jurisdiction over the whole of Bechuanaland, including the Tati District and the territory known as the disputed territory, excepting the area included in the Tuli District, and shall hold Courts at Kanye, Ramoutsa, Gaberone's, Mochudi, and Molepolole.

3. The Magistrate for Bechuanaland and the Tati District shall exercise jurisdiction within Khama's country, the disputed territory, and the Tati District, excepting the area included in the Tuli District, and shall hold Courts at Macloutsie, Palapye, and Tati.

* See vol. 1, page 183.

[Tati District, &c.]

4. The Magistrate at Tuli shall exercise jurisdiction within a 10 mile radius of Tuli and within the whole area comprised between the Shashi and Lundi Rivers.

5. The Magistrate at Fort Victoria shall exercise jurisdiction in the area comprised between the Lundi River and the parallel of Fort Charter.

6. The Magistrate at Fort Salisbury shall exercise jurisdiction in the area comprised between the parallel of Fort Charter and the Portuguese Possessions on the Zambesi, excepting the district placed under the Magistrate of Hartly Hill.

7. The Magistrate of Hartly Hill shall exercise jurisdiction in the area included in the Hartly Hill District, the Gold Fields of Lo Magundi, and the Lower Umfati.

8. The Magistrate at Umtali shall exercise jurisdiction in the Manica District between the Odzi River and the Portuguese Possessions.

9. Every Assistant Commissioner or Magistrate shall have and exercise such jurisdiction in all matters and causes, criminal and civil, as is had and exercised by the Courts of Resident Magistrates of the Colony of the Cape of Good Hope.

God save the Queen!

Given under my hand and seal, at Cape Town, this 27th day of June, 1891.

HENRY B. LOCH,
High Commissioner.

By command of his Excellency the High Commissioner.

GRAHAM BOWER,
Imperial Secretary.

App. 10.—*BRITISH NOTIFICATION.* Free Port of Zanzibar. *8th February,* 1892.

Foreign Office, February 8, 1892.*

Notice.

Free Port of Zanzibar.

Information has been received from Mr. Gerald Portal, C.B., Her Britannic Majesty's Agent and Consul-General at Zanzibar, that on the 1st instant he publicly declared that on and after that day import duties on all goods coming from foreign countries into the Port of Zanzibar would cease and be abolished.

The following articles are, however, for the public good, excepted from the terms of this declaration :—

1. Arms and munitions of war.
2. Alcoholic liquors, with the exception of beer and wines of lower strength than fifty degrees centigrade.
3. Kerosine and all other explosive oils or dangerous substances.

The duty on these latter articles will be remitted under certain conditions of storage.

All the above-mentioned articles will still remain subject to the duties leviable under existing treaties with foreign powers or under the provisions of the General Act of the Brussels Conference (No. 18), so soon as the latter shall come into force.

The above notice applies only to the Port of Zanzibar itself.

* "London Gazette," 9th February, 1892.

App. 11.—*NOTIFICATION to Treaty Powers. British Protectorate of Zanzibar placed under Free Trade Provisions of Berlin Act.* 22nd June, 1892.

Circular to Powers Signatories of Berlin Act.

MY LORD,
SIR, *Foreign Office, June* 22, 1892.

I HAVE to request you to notify to the Government to which you are accredited that it has been decided to place the British Protectorate of Zanzibar, from the 1st July next, under the free zone provisions of Article I of the Act of Berlin. (**No. 17.**)

The conditions under which the finances of Zanzibar were administered at the date of the passage of the Act were not consistent with the adoption of the fiscal system of the free zone, but under the Protectorate of Great Britain a complete change has been effected. The finances have been placed under European control, reforms have been introduced in every branch of the Administration, and sufficient progress has been made to justify Her Majesty's Government in notifying the acceptance of the invitation tendered by the Powers in 1885 to the Governments established on the African Littoral of the Indian Ocean.

Import Duties.

The whole of the Sultan's dominions, including the Islands of Zanzibar and Pemba, and the mainland territory under the administration of the Imperial British East Africa Company, will, from the above-named date, be placed permanently in the same financial position as that in which the Congo Free State was placed by the provisions of the Berlin Act (**No. 17**), afterwards modified by the Declaration annexed to the Brussels Act (vol. i, p. 88). The existing system under which the tariffs and duties are regulated by Commercial Treaties with individual Powers will be extinguished by the substitution for it of

[Free Trade.]

the system framed for the free zone by the assembled Powers in 1885.

Duties on Spirituous Liquors, Arms, Ammunition, and Explosives imported into Port of Zanzibar.

In making the above notification, your Excellency should explain that, although the stipulations of the Declaration annexed to the Act of Brussels will be applicable to the entire Protectorate, it is not proposed that the Sultan should avail himself at present, as regards the port of Zanzibar, of the right of levying import duties conferred by that Declaration. It has been decided that, until further notice, no such duties will be imposed in that port except upon spirituous liquors, arms, ammunition, and explosives.

5 Per Cent. Duty on Imports in other Zanzibar Ports administered by British East Africa Company. Benadir Ports, &c.

In all the other ports of Zanzibar, including those under the administration of the Imperial British East Africa Company and the Benadir ports, the 5 per cent. duty on imports now levied under Treaty will be replaced by a similar duty under the Declaration annexed to the Brussels Act (vol. i, p. 88). This will be in accordance with the terms of the Agreement respecting the tariff of the eastern zone of the Conventional Basin of the Congo, signed at Brussels on the 22nd December, 1890, by the Delegates of Great Britain, Germany, and Italy (**No, 19**). The tariff will be subject to the modifications as regards arms and ammunition, spirits, and certain specified articles, in accordance with the terms of the Agreement.

I am, &c.,
SALISBURY.

[Uganda.]

App. 12.—*PROVISIONAL AGREEMENT between King Mwanga, of Uganda, and Sir G. Portal.* 29th May, 1893.*

AGREEMENT between Mwanga, King of Uganda, and Sir Gerald Herbert Portal, Knight Commander of the Most Distinguished Order of St. Michael and St. George, a Companion of the Most Honourable Order of the Bath, Her Britannic Majesty's Commissioner and Consul-General for East Africa, &c.

1. Whereas the Imperial British East Africa Company have now definitely withdrawn from Uganda.

2. And whereas I, Mwanga, King of Uganda am profoundly and sincerely desirous of securing British protection for myself, my people, and dominions: as also assistance and guidance in the government of my country.

3. I, the said Mwanga, do hereby pledge and bind myself to the following conditions, with the object of securing the British protection, assistance, and guidance before mentioned:—

4. I undertake to make no Treaties or Agreements of any kind whatsoever with any Europeans of whatever nationality without the consent and approval of Her Majesty's Representative.

5. I freely recognise that so far as I, the King, am concerned, the sole jurisdiction over Europeans and over all persons not born in my dominions, and the settlement of all cases in which any such persons may be a party or parties, lie exclusively in the hands of Her Majesty's Representative.

6. In civil cases between my subjects the Court of Her Majesty's Representative shall be a Supreme Court of Appeal, but it shall lie entirely within the discretion of the said Representative to refuse to hear such appeals.

7. In criminal cases where only natives are concerned, it is left to the discretion of Her Majesty's Representative to interfere, in the public interest and for the sake of justice, to the extent and in the manner which he may consider desirable.

* Parliamentary Paper, "Africa, No. 2 (1894)," page 17.

8. And I, Mwanga, the King, undertake to see that due effect is given to all and every decision of the Court of Her Majesty's Representative under Articles 6 and 7.

9. I, Mwanga, fully recognise that the protection of Great Britain entails the complete recognition by myself, my Government, and people throughout my Kingdom of Uganda and its dependencies, of all and every international act and obligation to which Great Britain may be a party, as binding upon myself, my successors, and my said Government and people, to such extent and in such manner as may be prescribed by Her Majesty's Government.

10. No war or warlike operations of any kind shall be undertaken without the consent of Her Majesty's Representative, whose concurrence shall also be obtained in all serious matters of State, such as the appointment of Chiefs or officials, the political or religious distribution of territory, &c.

11. The assessment and collection of taxes, as also the disposal of the revenues of the country, are hereby made subject to the control and revision of Her Majesty's Government in such manner as they may from time to time direct.

12. The property of Her Majesty's Government and of their officers, and of all servants of Her Majesty's Government, shall be free from the incidence of all taxes.

13. Export and import duties on all goods leaving or entering Uganda and its dependencies shall be leviable by Her Majesty's Government for their sole use and benefit. These duties shall be fixed in accordance with the provisions of the General Acts of Berlin and Brussels of 1885 and 1890 (**Nos. 17 and 18**) respectively, and of any International Agreements arising from the same, and to which Great Britain is or may become a party.

14. The foreign relations of Uganda and its dependencies are hereby placed unreservedly in the hands of Her Majesty's Representative.

15. Slave trading or slave raiding, or the exportation or importation of people for sale or exchange as slaves, is prohibited. I, Mwanga, also undertake, for myself and my successors, to give due effect to such laws and regulations, having

for their object the complete ultimate abolition of the status of slavery in Uganda and its dependencies, as may be dictated by Her Majesty's Government.

16. In consideration of the above engagements on the part of Mwanga, King of Uganda, I, Gerald Herbert Portal, K.C.M.G., C.B., Her Britannic Majesty's Commissioner and Consul-General for East Africa, on behalf of Her Majesty's Government, do hereby agree to appoint and leave a British Representative with a sufficient staff to carry out the provisions of this Agreement, which is entirely subject to the approval and ratification of Her Majesty's Government,* and is therefore only binding until such time as the decision of Her Majesty's Government can be conveyed to, and reach Uganda. In the event of Her Majesty's Government being willing to assent to the above conditions and terms, Mwanga, the King, undertakes hereby, on behalf of himself and his successors, to make a Treaty in the above or a similar sense either in perpetuity or for such specified period as Her Majesty's Government may desire.

17. The present Agreement supersedes all other Agreements or Treaties whatsoever made by Mwanga or his predecessors.

18. This Agreement shall come into force from the date of its signature.

In faith whereof we have respectively signed this Agreement, and have thereunto affixed our seals.

Done in duplicate at Kampala, this 29th of May, A.D. 1893.

KABAKA (King).
G. H. PORTAL.

Witnesses to the signatures of King Mwanga and Sir Gerald Portal;

ERNEST J. L. BERKELEY.
KATIKIRO APOLLO.

Kampala, May 29, 1893.

* See Notification. British Protectorate over Uganda, 18th June, 1894, page 1016.

App. 13. —*FURTHER NOTES on Dahomey. Jan.—June,*
1894.

On the 5th January, 1894, a Declaration was signed by General Dobbs, accepting the submission of the Princes, Chiefs, and inhabitants of Dahomey, and placing the country on the left bank of the Ouemé under French Protection.

The Kingdom of Dahomey is now divided into two States, having for their capitals Abomey and Allada.

On the 15th January, 1894, Ago-il-Agbo, son of Gléglé, was named King of Abomey, and on the 29th of the same month the new Sovereign concluded a Treaty with General Dobbs recognising the French Protectorate.

On the 4th February, 1894, the new King of Allada, Gi-Gla-Uonon, was installed as Sovereign of the southern portion of the kingdom, and also placed under the Protectorate of France.

On the 22nd June, 1894, a Presidential Decree was published in the "Journal Officiel" organising the coast territory of Dahomey as a French Colony, entitled "Dahomey et dépendances."

[Cameroons and French Congo. Lake Tchad.]

App. 14.—*PROTOCOL between the French and German Delegates for the Settlement of the Questions pending between the Two Countries in the Region comprised between the Colonies of the Cameroons and French Congo; and to fix the Line of Demarkation of their respective Spheres of Influence in the Region of Lake Tchad. Berlin, February 4th*, 1894.*

PROTOCOLE.

Les Soussignés:

Docteur Paul Kayser, Conseiller privé actuel de Légation, Dirigeant les Affaires Coloniales au Département des Affaires Etrangères;

Docteur Alexandre Baron de Danckelman, Professeur;

Jacques Haussmann, Chef de Division au Sous-Secrétariat d'Etat des Colonies;

Parfait-Louis Monteil, Chef de Bataillon d'Infanterie de Marine;

Délégués par le Gouvernement de l'Empire Allemand et par le Gouvernement de la République Française à l'effet de préparer un accord destiné à régler les questions pendantes entre l'Allemagne et la France dans la région comprise entre les Colonies du Cameroun et du Congo Français et à établir la ligne de démarcation des zônes d'influence respectives des deux Pays dans la région du Lac Tchad, sont convenus des dispositions suivantes:

ART. I. La frontière entre la Colonie du Cameroun et la Colonie du Congo Français suivera, à partir de l'intersection du parallèle formant la frontière avec le méridien 15° Greenwich (12° 40' Paris), le dit méridien jusqu'à sa rencontre avec la Rivière Ngoko; le Ngoko jusqu'à sa rencontre avec le parallèle 2°;† de là, en se dirigeant vers l'Est, ce parallèle jusqu'à sa rencontre avec la Rivière Sangha. Elle suivra ensuite, en remontant vers le Nord, sur une longuer de 30 kilomètres, la Rivière Sangha; du point qui sera ainsi déterminé sur la Rive droite de la Sangha, une ligne droite aboutissant sur le paral-

* "Deutschen Kolonialblatts (Extra-Nummer)," 10th March, 1894.
† Voir Annexe, § II, p. 983.

lèle de Bania, à soixante-deux minutes (62′) à l'Ouest de Bania, de ce point, une ligne droite aboutissant, sur le parallèle de Gaza, à quarante-trois minutes (43′) à l'Ouest de Gaza.

De là, la frontière se dirigera en ligne droite vers Koundé, laissant Koundé à l'Est avec une banlieue déterminée à l'Ouest par un arc-de-cercle d'un rayon de 5 kilomètres, partant, au Sud, du point où il sera coupé par la ligne allant à Koundé, et finissant au Nord, à son intersection avec le méridien de Koundé ; de là, la frontière suivra le parallèle de ce point jusqu'à sa rencontre avec le méridien 15° Greenwich (12° 40′ Paris).*

Le tracé suivra ensuite le méridien 15° Greenwich (12° 40′ Paris) jusqu'à sa rencontre avec le parallèle 8° 30′, puis, une ligne droite aboutissant à Lamé, en laissant une banlieue de 5 kilomètres à, l'Ouest de ce point ; de Lamé, une ligne droite aboutissant sur la rive gauche du Mayo-Kebbi, à hauteur de Bifara.† Du point d'accès à la rive gauche du Mayo-Kebbi, la frontière traversera la rivière et remontera en ligne droite vers le Nord, laissant Bifara à l'Est, jusqu'à la rencontre du 10ᵉ parallèle.—Elle suivra ce parallèle jusqu'à sa rencontre avec le Chari,‡ enfin le cours du Chari jusqu'au Lac Tchad.§

Art. II. Le Gouvernement allemand et le Gouvernement français prennent l'engagement réciproque de n'exercer aucune action politique dans les sphères d'influence qu'ils se reconnaissent par la ligne de démarcation déterminée à l'article précédent. Il est convenu par là que chacune des deux Puissances s'interdit de faire des acquisitions territoriales, de conclure des traités, d'accepter des droits de souveraineté ou de protectorat, de gêner ou de contester l'influence de l'autre Puissance dans la zône qui lui est réservée.

Art. III. L'Allemagne, en ce qui concerne la partie des eaux de la Bénoué et de ses affluents comprise dans sa sphère d'influence ; la France, en ce qui concerne la partie du Mayo-Kebbi et des autres affluents de la Bénoué comprise dans sa sphère d'influence se reconnaissent respectivement tenues d'appliquer et de faire respecter les dispositions relatives à la

* See Annexe, § III, p. 983.
† See Annexe, § IV, p. 983.
‡ See Annexe, § III, p. 983.
§ See Annexe, § V, p. 983.

Map to illustrate protocol between
FRANCE AND GERMANY
of 4th February, 1894.

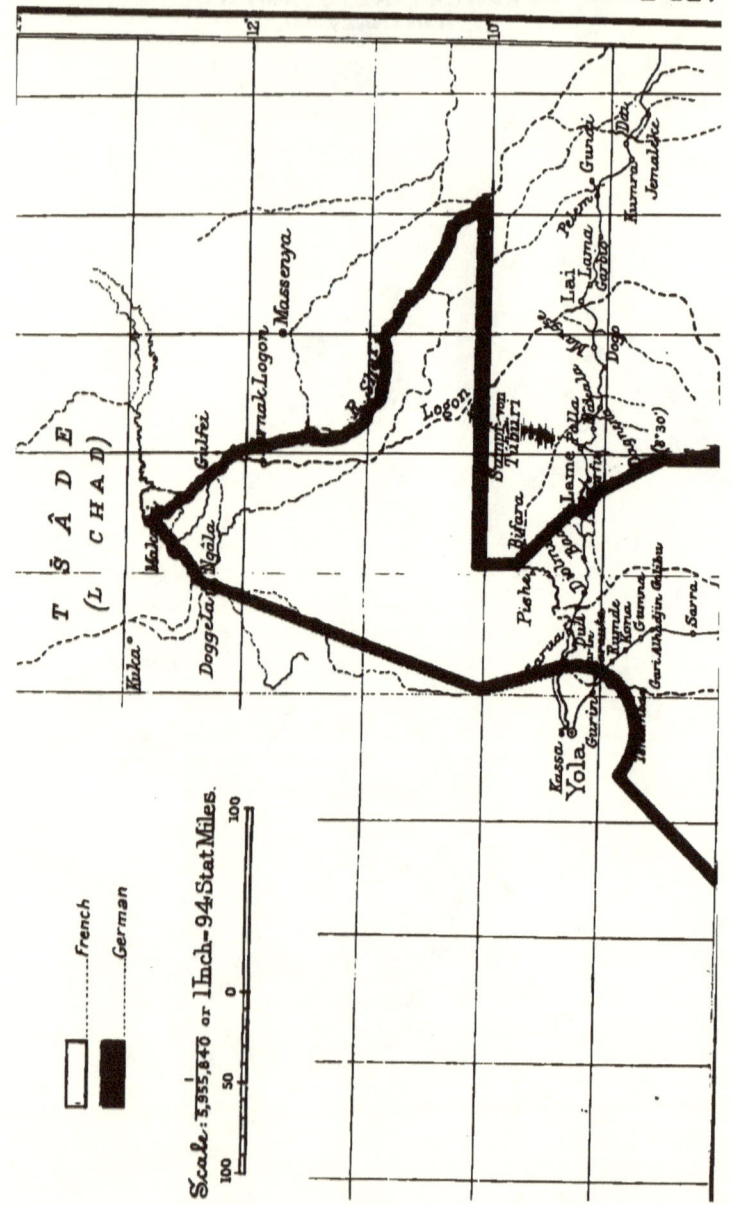

To face page

FRANCE AND GERMANY OF 4TH FEBRUARY 1894.

Map to illustrate protocol between
FRANCE AND GERMANY
of 4th February, 1894.

[Boundaries. Cameroons, French Congo, and Lake Tchad.]

liberté de navigation et de commerce énumérées dans les Articles XXVI, XXVII, XXVIII, XXIX, XXXI, XXXII, XXXIII de l'Acte de Berlin du 26 février 1885 (**No. 17**), de même que les clauses de l'Acte de Bruxelles relatives à l'importation des armes et des spiritueux. (**No. 18**.)

L'Allemagne et la France s'assurent respectivement le bénéfice de ces mêmes dispositions en ce qui concerne la navigation du Chari, du Logone et de leurs affluents et l'importation des armes et des spiritueux dans les bassins de ces rivières.

ART. IV. Dans les territoires de leurs zônes d'influence respectives compris dans les bassins de la Bénoué et de ses affluents, du Chari, du Logone et de leurs affluents, de même que dans les territoires situés au Sud et au Sud-Est du Lac Tchad, les commerçants ou les voyageurs des deux pays seront traités sur le pied d'une parfaite égalité en ce qui concerne l'usage des routes ou autres voies de communication terrestres. Dans ces mêmes territoires, les Nationaux des deux Pays seront soumis aux mêmes règles et jouiront des mêmes avantages au point de vue des acquisitions et installations nécessaires à l'exercice et au développement de leur commerce et de leur industrie.

Sont exclus de ces dispositions les routes et voies terrestres de communication des bassins côtiers de la Colonie du Cameroun, ou des bassins côtiers de la Colonie du Congo Français non compris dans le bassin conventionnel du Congo tel qu'il a été défini par l'Acte de Berlin. (**No. 17.**)

Ces dispositions, toutefois, s'appliquent à la route Yola, Ngaoundéré, Koundé, Gaza, Bania et vice-versa, telle qu'elle est repérée sur la carte annexée au présent protocole, alors même qu'elle serait coupée par des affluents des bassins côtiers.

Les tariffs des taxes ou droits qui pourront être établis de part et d'autre ne comporteront, à l'égard des commerçants des deux pays, aucun traitement différentiel.

ART. V. En foi de quoi les Délégués ont dressé le présent protocole et y ont apposé leur signature.

Fait à Berlin, en double expédition, le 4 Février 1894.

KAYSER.   HAUSSMANN.
VON DANCKELMAN.   MONTEIL.

[Boundaries. Cameroons, French Congo, and Lake Tchad.]

ANNEXE.

§ I.—La ligne de démarcation des sphères d'influence respectives des deux Puissances contractantes telle qu'elle est décrite à l'article 1ᵉʳ du protocole du même jour (p. 980) sera conforme au tracé porté sur la carte annexée au présent protocole qui a été établie d'après les données géographiques actuellement connues et admises de part et d'autre.

§ II —Dans le cas où la rivière Ngoko, à partir de son intersection avec le méridien 15° Greenwich (12° 40′ Paris) ne couperait pas le 2ᵉ parallèle, la frontière suivrait le Ngoko sur une longueur de 35 kilomètres à l'Est de son intersection avec le mériden 15° Gr. (12° 40′ Paris) ; à partir du point ainsi déterminé à l'Est, elle rejoindrait par une ligne droite l'intersection du 2ᵉ parallèle avec la Sangha.

§ III.—S'il venait à être démontré à la suite d'observations nouvelles dûment vérifiées, que les positions de Bania, de Gaza ou de Koundé sont erronées, et que, par suite la frontière telle qu'elle est définie par le présent protocole, se trouve reportée, au regard de l'un de ces trois points, d'une distance supérieure à dix minutes de degré (10′) à l'Ouest du méridien 15° Greenwich (12° 40′ Paris), les deux Gouvernements se mettraient d'accord pour procéder à une rectification du tracé, de manière à établir une compensation équivalente au profit de l'Allemagne dans la région en question.

Une rectification du même genre interviendrait, en vue d'établir une compensation au profit de la France, s'il était démontré que l'intersection du parallèle 10ᵉ avec le Chari reporte la frontière à une distance de plus de dix minutes (10′) à l'Est du point indiqué sur la carte (Longitude 17° 10′ Greenwich—14° 50′ Paris).

§ IV.—En ce qui concerne le point d'accès au Mayo-Kebbi, il demeure entendu que, quelle que soit la position définitivement reconnue pour ce point, la frontière laissera dans la sphère d'influence française les villages de Bifara et de Lamé.

§ V.—Dans le cas ou le Chari, depuis Goulfeï jusqu'à son embouchure dans le Tchad, se diviserait en plusieurs bras, la frontière suivrait la principale branche navigable jusqu'à l'entrée

App. 14]   FRANCE AND GERMANY.   [4 Feb., 1894.

[Boundaries. Cameroons, French Congo, and Lake Tchad.]

dans le Tchad, avec cette réserve que, pour que ce tracé soit définitif, la différence de longitude entre le point ainsi atteint par la frontière sur la Rive Sud du Tchad et Kouka, capitale du Bornou, pris comme point fixe, sera de un degré. Dans le cas où des observations ultérieures, dûment vérifiées, démontreraient que l'écart en longitude entre Kouka et la dite embouchure diffère de cinq minutes de degré (5'), en plus ou en moins, de celui qui vient d'être indiqué, il y aurait lieu, par une entente amiable, de modifier le tracé de cette partie de la frontière de manière que les deux pays conservent, au point de vue de l'accès au Tchad, et des territoires qui leur sont reconnus dans cette région, des avantages équivalents à ceux qui leur sont assurés par le tracé porté sur la carte annexée au présent protocole.

§ VI.—Toutes les fois que le cours d'un fleuve ou d'une rivière est indiqué comme formant la ligne de démarcation, c'est le thalweg du fleuve ou de la rivière qui est considéré comme frontière.

§ VII.—Les deux Gouvernements admettent qu'il y aura lieu, dans l'avenir, de substituer progressivement aux lignes idéales qui ont servi à déterminer la frontière telle qu'elle est définie par le présent protocole un tracé déterminé par la configuration naturelle du terrain et jalonné par des points exactement reconnus, en ayant soin, dans les accords qui interviendront à cet effet, de ne pas avantager l'une des deux Parties sans compensation équitable pour l'autre.

Vu pour être annexé au protocole du 4 Février, 1894.

———

On the 15th March, 1894, a Convention was concluded at Berlin between the German and French Governments, confirming the above Protocol (see "Journal Officiel," 14th August, 1894).

App. 15.—*DECLARATION. Congo and Portugal. Approval of Report of Boundary Commissioners of 26th June, 1893. Lunda Region. Brussels, 24th March, 1894.*

(Translation.)

DECLARATION signed at Brussels, 24th March, 1894, conveying the approval by the Governments of the Independent State of the Congo and of His Most Faithful Majesty of the tracing of the frontier executed by their Commissioners in the region of Lunda, in execution of the Convention concluded at Lisbon 25th May, 1891 (**No. 59**).

Declaration.

The Governments of the Independent State of the Congo and of His Most Faithful Majesty, having received the report of the delimitation works carried out on the spot by the Commissioners charged by them, in the terms of Article II of the Convention signed at Lisbon, 25th May, 1891 (**No. 59**), to execute the tracing of the boundary in accordance with Article I of the above-mentioned Convention, and having taken cognisance of the procès-verbal of the 26th June, 1893, signed, subject to ratification, at Loanda, have decided to approve and ratify respectively this procès-verbal of the 26th June, 1893, in the following terms :—

The year eighteen hundred and ninety-three, the twenty-sixth day of the month of June,

We, George Grenfell, missionary of the English Baptist Mission, and Jayme Lobo de Brito Godins, Governor General *ad interim* of the province of Angola;

Having exchanged our diplomas, found in good and due form, giving us full powers as Royal Commissioners for the Independent State of the Congo and for Portugal to execute conjointly the tracing of the boundary in the region of Lunda, while complying with the stipulations laid down in Articles I· and II of the Convention of Lisbon of the 25th May, 1891 (**No. 59**); the Royal Commissioner for Portugal having in addition the right of transferring wholly or in part the powers

App 15] CONGO AND PORTUGAL. [24 March, 1894.

[Boundary. Lunda Region.]

which have been conferred upon him, which faculty he has made use of by delegating his powers to Simão-Candido Sarmento, Lieutenant Graduate of the Portuguese Army, in so far as they relate to the works on the spot.

Having taken cognisance of the annexed procès-verbaux of the five sittings, which are signed by the aforesaid George Grenfell, Royal Commissioner, and Lieutenant Simão-Candido Sarmento, delegate of the Royal Portuguese Commissioner for the works on the spot, and also by the Captain in command of the public force of the Independent State of the Congo, Florent Gorin, Royal Commissioner for the technical works, we decide to adopt *ad referendum* the tracing of the boundary, set forth in the present Act, which shall not be signed by the aforesaid Captain in command, Florent Gorin, who happens to be absent, which fact shall not lessen the value of the present document, in that it is the transcription of the boundaries that the above-mentioned Captain in command, Florent Gorin, has approved, which are mentioned in the five procès-verbaux aforesaid.

Following the thalweg of the Kwango (Cuango) from the 8th parallel as far as its confluence with the Tungila (Utunguila) 8° 7' 40" south latitude approx.; the thalweg of the Tungila (Utunguila) as far as its intersection with the canal through which pass the waters of the Lola; the thalweg of the same canal as far as its junction with the Komba, 8' west of the Wamba (Uhamba), and 8° 5' 4" south latitude approx.; for want of a natural boundary, the frontier as far as the thalweg of the Wamba (Uhamba) shall be marked out by the line due east, passing through the aforesaid point of junction (Komba and Lola).

The thalweg of the Wamba (Uhamba) from the parallel of the point of junction between the Komba (Comba) and the Lola, as far as its confluence with the Uövo (Nuovo); the thalweg of the Uövo (Nuovo) as far as its junction with the N'Kombo (Combo); the thalweg of the N'Kombo and of the Kamanguna (Camanguna) (or the river by which the waters of the river Lué flow into the N'Kombo), as far as the 8th degree south latitude. From this point the boundary shall be the 8th parallel, as far as the thalweg of the Lucaïa,

[Boundary. Lunda Region.]

then the thalweg of this river (Lukaï) as far as 7° 55' south latitude; the parallel from this point (7°.55' south latitude) as far as the Kwengo (Cuengo); from this point the thalweg of the Kwengo (Cuengo), as far as the 8th degree; from thence a parallel as far as the river Luita; the thalweg of the Luita as far as its junction with the Kivilu (Cuilu). From thence (7° 34' south latitude approx.) the parallel as far as the thalweg of the Kama Bomba (Camabomba) or Kangulungu (Congulungu); the thalweg of the Kangulungu as far as the junction of its waters with the Loangué, and the thalweg of the Loangué as far as 7° south latitude. From the intersection of the thalweg of the Loangué and of the 7th degree, following this parallel as far as its intersection with the thalweg of the Lovua; the thalweg of the Lovua as far as 6° 55' south latitude. From this point (6° 55' south latitude) the boundary shall be marked out by the parallel as far as its intersection with the thalweg of the Chikapa (Chicapa); the thalweg of this river (Chicapa) as far as 7° 17' south latitude; from this point (7° 17' south latitude) the parallel as far as the thalweg of the Kassaï (Cassai).

Done at Loanda, in duplicate original, the twenty-sixth day of the month of June of the year Eighteen hundred and ninety-three.

    For the Independent State of the Congo,
 (Signed) GEORGE GRENFELL,

    For Portugal,
 (Signed) JAYME LOBO DE BRITO GODINS.

To this effect the undersigned, His Excellency Count de Grelle-Rogier, Secretary of State for Foreign Affairs of the Independent State of the Congo, and His Excellency Monsieur M. Martins d'Antas, Envoy Extraordinary and Minister Plenipotentiary of His Most Faithful Majesty, duly authorised, have embodied in the present declaration the ratification by their respective Governments of the preceding Act, the said ratification to come into full and entire force on the date of the thirty-first of March, Eighteen hundred and ninety-four.

CONGO AND PORTUGAL. [24 March, 1894.

[Boundary. Lunda Region.]

In witness whereof the undersigned have drawn up the present declaration, which they have signed in duplicate, and to which they have affixed their seals.

Done at Brussels, the twenty-fourth day of the month of March, Eighteen hundred and ninety-four.

> The Plenipotentiary of His Majesty the Sovereign King of the Independent State of the Congo,
> C^{TE} DE GRELLE-ROGIER.
>
> The Plenipotentiary of His Most Faithful Majesty,
> MIGUEL MARTINS D'ANTAS.

12 May, 1894.] CONGO AND GREAT BRITAIN. [App. 16

[Spheres of Influence in East-and Central Africa.]

App. 16.—*AGREEMENT between Great Britain and His Majesty King Leopold II, Sovereign of the Independent State of the Congo, relating to the Spheres of Influence of Great Britain and the Independent State of the Congo in East and Central Africa. Signed at Brussels, 12th May, 1894.**

THE Undersigned, the Honourable Sir Francis Richard Plunkett, a Knight Grand Cross of the most distinguished Order of St. Michael and St. George, Her Britannic Majesty's Envoy Extraordinary and Minister Plenipotentiary to the King of the Belgians, on behalf of the British Government, and M. van Eetvelde, Officer of the Order of Leopold, Grand Cross of the Orders of St. Gregory the Great, of Christ of Portugal, and of the African Redemption, &c., Secretary of State of the Interior of the Independent State of the Congo, on behalf of the Government of the Independent State of the Congo, duly authorised by their respective Governments, have agreed as follows:

His Majesty the King of the Belgians, Sovereign of the Independent State of the Congo, having recognised the British sphere of influence, as laid down in the Anglo-German Agreement of the 1st July, 1890 (**No. 129**), Great Britain undertakes to give to His Majesty a lease of territories in the western basin of the Nile, under the conditions specified in the following Articles:

Boundary. North of German Sphere. Watersheds between the Nile and the Congo.

ART. I.—(*a.*) It is agreed that the sphere of influence of the Independent Congo State shall be limited to the north of the German sphere in East Africa by a frontier following the 30th meridian east of Greenwich up to its intersection by the watershed between the Nile and the Congo, and thence following this watershed in a northerly and north-westerly direction.

* Parliamentary Paper, "Treaty Series No. 15 (1894)."

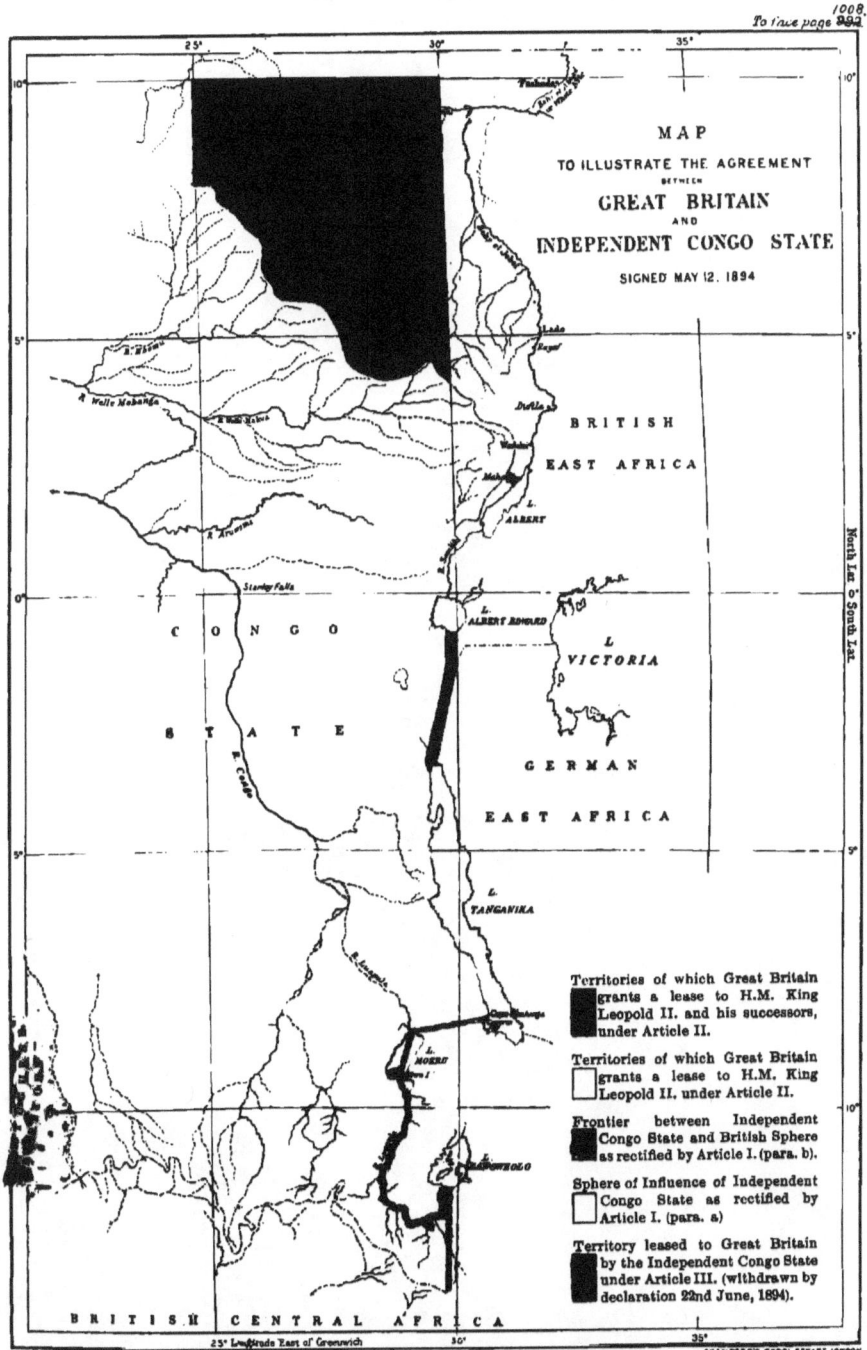

Map to illustrate the agreement between
GREAT BRITAIN & THE CONGO FREE STATE
of 12th of May, 1894.

Boundary. North of the Zambesi. Luapula River. Lake Moero to Lake Bangweolo.

(*b.*) The frontier between the Independent Congo State and the British sphere to the north of the Zambesi shall follow a line running direct from the extremity of Cape Akalunga on Lake Tanganika, situated at the northernmost point of Cameron Bay at about 8° 15' south latitude, to the right bank of the River Luapula, where this river issues from Lake Moero. The line shall then be drawn directly to the entrance of the river into the lake, being, however, deflected towards the south of the lake so as to give the Island of Kilwa to Great Britain. It shall then follow the "thalweg" of the Luapula up to its issue from Lake Bangweolo. Thence it shall run southwards along the meridian of longitude of the point where the river leaves the lake to the watershed between the Congo and Zambesi, which it shall follow until it reaches the Portuguese frontier.

Lease of certain Territories by Great Britain to the Congo State. West Shore of Lake Albert and Watershed between the Nile and the Congo.

ART. II. Great Britain grants a lease to His Majesty King Leopold II, Sovereign of the Independent Congo State, of the territories hereinafter defined, to be by him occupied and administered on the conditions and for the period of time hereafter laid down.

Boundaries.

The territories shall be bounded by a line starting from a point situated on the west shore of Lake Albert, immediately to the south of Mahagi, to the nearest point of the frontier defined in paragraph (*a*) of the preceding Article. Thence it shall follow the watershed between the Congo and the Nile up to the 25th meridian east of Greenwich, and that meridian up to its intersection by the 10th parallel north, whence it shall run along that parallel directly to a point to be determined to the north of Fashoda. Thence it shall follow the "thalweg" of the Nile southward to Lake Albert, and the western shore of Lake Albert to the point above indicated south of Mahagi.

[Spheres of Influence in East and Central Africa.]

This lease shall remain in force during the reign of His Majesty Leopold II, Sovereign of the Independent Congo State.

Nevertheless, at the expiration of His Majesty's reign, it shall remain fully in force as far as concerns all the portion of the territories above mentioned situated to the west of the 30th meridian east of Greenwich, as well as a strip of 25 kilom. in breadth, to be delimitated by common consent, stretching from the watershed between the Nile and the Congo up to the western shore of Lake Albert, and including the port of Mahagi.

This extended lease shall be continued so long as the Congo territories as an Independent State or as a Belgian Colony remain under the sovereignty of His Majesty and His Majesty's successors.

Flag.

Throughout the continuance of a lease there shall be used a special flag in the leased territories.

Lease of Territory by Congo State to Great Britain between Lake Tanganika and Lake Albert Edward.

[ART. III.* The Independent Congo State grants under lease to Great Britain, to be administered when occupied, under the conditions and for a period hereafter determined, a strip of territory 25 kilom. in breadth, extending from the most northerly port on Lake Tanganika, which is included in it, to the most southerly point of Lake Albert Edward.

This lease will have similar duration to that which applies to the territories to the west of the 30th meridian east of Greenwich.]

Self-Denying Declaration.

ART. IV. His Majesty King Leopold II, Sovereign of the Independent Congo State, recognizes that he neither has nor seeks to acquire any political rights in the territories ceded to him under lease in the Nile Basin other than those which are in conformity with the present Agreement.

Similarly, Great Britain recognizes that she neither has, nor

* This Article was withdrawn by a Declaration, signed 22nd June, 1894. See p. 1017.

seeks to acquire, any political rights in the strip of territory granted to her on lease between Lake Tanganika and Lake Albert Edward other than those which are in conformity with the present Agreement.

Telegraphic Communication.

ART. V. The Independent Congo State authorizes the construction through its territories by Great Britain, or by any Company duly authorized by the British Government, of a line of telegraph connecting the British territories in South Africa with the British sphere of influence on the Nile. The Government of the Congo State shall have facilities for connecting this line with its own telegraphic system.

This authorization shall not confer on Great Britain or any Company, person or persons, delegated to construct the telegraph line, any rights of police or administration within the territory of the Congo State.

Equality of Treatment in Territories Leased.

ART. VI. In the territories under lease in this Agreement the subjects of each of the Contracting Parties shall reciprocally enjoy equal rights and immunities, and shall not be subjected to any differential treatment of any kind.

In witness whereof the Undersigned have signed the present Agreement, and have affixed thereto the seal of their arms.

Done in duplicate at Brussels, this 12th day of May, 1894.

    (L.S.) FRANCIS RICHARD PLUNKETT.
    (L.S.) EDM. VAN EETVELDE.

Claims of Turkey and Egypt in Basin of the Upper Nile not Ignored.

(1.) *Sir F. Plunkett to M. van Eetvelde.*

British Legation, Brussels.
M. le Secrétaire d'État,   *May* 12, 1894.

THE Earl of Kimberley, in authorizing me to sign the

12 May, 1894.] CONGO AND GREAT BRITAIN. [App. 16.

[Spheres of Influence in East and Central Africa.]

Agreement of this day's date for a lease of certain territories in the British sphere of influence in East Africa to His Majesty King Leopold II, has directed me to record the assurance that the parties to the Agreement do not ignore the claims of Turkey and Egypt in the basin of the Upper Nile.

I avail, &c.,
F. R. PLUNKETT.

(2.) *M. van Eetvelde to Sir F. Plunkett.*

Brussels,
Sir, May 12, 1894.

IN signing, on behalf of His Majesty Leopold II, the Agreement of this day's date, for a lease of certain territories in the British sphere of influence in East Africa, I reciprocate the assurance that the parties to the Agreement do not ignore the claims of Turkey and Egypt in the basin of the Upper Nile.

I avail, &c.,
EDM. van EETVELDE.

Recruitment of Soldiers by British Authorities.

(3.) *M. van Eetvelde to Sir F. Plunkett.*

(Translation.)

M. le Ministre, Brussels, May 12, 1894.

IN the course of the discussions to which the Convention of to-day between the Independent State of the Congo and Great Britain has given rise, I have had occasion to declare to you that the State of the Congo engages to authorize, in case of need, such recruitment of soldiers as the Agents duly commissioned for that purpose by the British authorities may wish to effect in the territories situated between the 30th meridian and Lake Albert.

I have the honour to confirm this engagement, and I seize, &c.

EDM. van EETVELDE.

1012

App. 16]   CONGO AND GREAT BRITAIN.   [12 May, 1894.

[Spheres of Influence in East and Central Africa.]

Recruitment in British Colonies on West Coast of Africa for service in Western Basin of the Nile.

4.—*Sir F. Plunkett to M. van Eetvelde.*

British Legation, Brussels,
M. le Secrétaire d'État,   May 12, 1894.

IN accordance with the wish which you have expressed, I have to convey to your Excellency the assurance, on the part of the Earl of Kimberley, that his Lordship will be ready to recommend to Her Majesty's Secretary of State for the Colonies that facilities shall be given, so far as it may be found to be practicable, for recruitment, under suitable conditions, in the British Colonies on the West Coast of Africa, to facilitate the prompt and complete occupation by His Majesty King Leopold II of the territories in the western basin of the Nile comprised in the lease contained in the Agreement of this day's date.

I avail, &c.,
F. R. PLUNKETT.

Explanatory Despatch relating to the above Agreement between Great Britain and the Congo State, of 12th May, 1894.*

British Sphere of Influence.   Uganda, &c.

The Earl of Kimberley to Mr. Hardinge.

Sir,   Foreign Office, May 23, 1894.

WHEN Her Majesty's Government decided upon assuming the Protectorate of Uganda, it became incumbent on them to consider the position of Great Britain as regards that part of the British sphere described in the Anglo-German Agreement as the western watershed of the Nile. It was understood that in 1890 arrangements were made between the Administrator of the Congo Free State and the late Sir W. Mackinnon, under which the East Africa Company agreed to waive in favour of the Free State any powers which it might acquire in the territory so described as a Chartered Company administering in the British sphere with the sanction of the Crown. The documents

* Parl. Paper, " Africa, No. 4 (1894)."

1013

recording whatever arrangements may have been conluded were not officially communicated to, nor sanctioned by Her Majesty's Government, and obviously could not have validity without that sanction. The Free State' Administration, however, appears to have considered that, in virtue of these arrangements, it was justified in sending exploring parties into the territory affected by them. The expeditions are believed to have travelled over a considerable portion of the territory, and it appears that their leaders made Treaties and established posts.

Her Majesty's Minister at Brussels was from time to time, directed to point out that, though Her Majesty's Government had no accurate information as to the destination and proceedings of these expeditions, the territory thus explored was well known to be included in the British sphere of influence.

Her Majesty's Government, in examining this situation in connection with the Protectorate of Uganda, desired, in order to put an end to all controversy as to these proceedings, to arrive at an arrangement which would be satisfactory to both parties. They could not fail to recognize the sacrifices which had been made in endeavouring to open up the country by His Majesty the King of the Belgians, whose efforts to promote the civilization of Africa have commanded their warm sympathy.

Claims of Egypt and Turkey to Equatorial Provinces.

On approaching His Majesty they found him fully disposed to enter into an arrangement which, while enabling him to continue the work he had commenced, would record his recognition of the position of Great Britain in her sphere, and of such claims as Egypt, and, through her, Turkey, may have to the Equatorial Provinces whose administration was abandoned owing to the evacuation of the Soudan.

I enclose copy of an Agreement by which His Majesty having recognized, on behalf of the Congo State, the British sphere of influence as laid down in the Anglo-German Agreement of 1890 (**No. 129**), received from Great Britain leases of the territory specified in the Agreement under certain conditions.

Her Majesty's Government are satisfied that, under the Agreement, this portion of the British sphere will be adminis-

[Spheres of Influence in East and Central Africa.]

tered in a spirit in full accordance with the requirements of civilization, and of the Acts of Berlin and Brussels (**Nos. 17, 18**).

The Agreement also effects certain frontier rectifications with the Congo Free State, which remove causes of possible local friction, and adds to the delimitations already concluded with Italy on the north (**Nos. 135, 136**) and Germany on the south (**No. 129**), delimitation between the British sphere and the conterminous Power on the west.

Finally, Article 3 provides for the lease to Great Britain of a port at the northern end of Lake Tanganyika.* As the southern end of the lake is within the British sphere in Central Africa, this concession will materially facilitate communication between the two British spheres. In order to secure access to this port, the lease has been obtained of a road passing through the Free State territory, connecting Lake Albert Edward, the eastern shore of which is in the British sphere, with the leased port. The navigation of Lake Tanganyika being declared to be free by the IInd Article of the Act of Berlin of 1885 (**No. 17**), this arrangement secures to British trade uninterrupted communication, the value of which is completed by the guarantees as to freedom of transit recorded in the IVth Article of the Berlin Act (**No. 17**), and the VIIIth Article of the Anglo-German Agreement of 1890 (**No. 129**).

A map is annexed showing the effect of the Agreement.

I have, &c.,
KIMBERLEY.

* This Article was withdrawn by a Declaration, signed 22nd June, 1894. See p. 1017.

App. 17.—*NOTIFICATION. British Protectorate over Uganda. London, 18th June, 1894.**

Foreign Office, June 18, 1894.

It is hereby notified, for public information, that under and by virtue of the agreement concluded on the 29th May, 1893 (page 995), between the late Sir G. Portal and Mwanga, King of Uganda, the country of that ruler is placed under the Protectorate of Her Majesty the Queen.

This Protectorate comprises the territory known as Uganda proper, bounded by the territories known as Usoga, Unyoro, Ankoli, and Koki.

* "London Gazette," 19th June, 1894.

App. 18.—*DECLARATION between Great Britain and the Congo Free State, withdrawing Art. III of the Agreement of 12th May, 1894, respecting the Territory between Lake Tanganika and Lake Albert Edward. Brussels, 22nd June, 1894.**

Declaration.

IN compliance with the request made by His Majesty the King of the Belgians, Sovereign of the Independent State of the Congo, that the Government of Her Britannic Majesty will consent to the withdrawal of Art. III of the Agreement of the 12th May, 1894 (p. 1008), the Undersigned, duly authorised by their respective Governments, agree that the said Article be withdrawn.

Done, in duplicate, at Brussels, the 22nd day of June, 1894.

F. R. PLUKETT.
EDMOND VAN EETVELDE.

* Parl. Paper, "Africa, No. 5 (1894)," p. 5, and "Treaty Series No. 20 (1894)."

App. 19.—*TREATY. Royal Niger Company and Gandu. Jurisdiction over Foreigners, &c.* 4*th July*, 1894.

In the name of the most merciful God !

TREATY made on July 4, 1894, between Omoru, Sultan of Gandu, on the one hand, and Mr. Wallace, on the other hand, for and on behalf of the Royal Niger Company (Chartered and Limited).

I, THE Undersigned Omoru, Sultan of Gandu, hereby confirm the Treaties made between the Sultan Maleki, whom I succeeded, and Thomson and King, on behalf of the Royal Niger Company (Chartered and Limited), the latter Treaty made on the [7th] day of April, 1890 (page 983). I now confirm these Treaties.

2. With my own hand I bind myself with Wallace, on behalf of the Company, and accept this following Treaty, made on the 4th day of July, 1894.

3. With the view of bettering the condition of my people, and having considered and taken counsel with my Chiefs, 1 give to the Company and their successors for ever full power and rights in perpetuity over foreigners in my country, whether travelling or resident, including right of just taxation as they may see fit. My Chiefs are in no way to interfere, and are to recognize no one but the Company.

4. I give to the Company and their successors for ever all power in any part of my dominions as to mining rights.

5. The Company bind themselves not to exercise any monopoly of trade.

6. I recognize that the Company received their power from the Queen of Great Britain, and that they are Her Majesty's Representatives to me. I will not recognize any other white nation, because the Company are my help.

7. I state that the country of Illorin and the country of Gurma are included in my dominions, the latter extending to Libtako.

App. 19]   GREAT BRITAIN (NIGER).   [4 July, 1894.

[Gandu.]

8. The Company undertake not to interfere with the customs of the Mussulmans, but to maintain friendly relations.

9. In recognition of the Treaties between us, the Company have paid me a subsidy of 2,000 bags, as hitherto annually for the past nine years. They have acted honourably towards me, and as I should desire.

I hereby confirm the previous Treaties, and accept this Treaty for myself, for my heirs and for my successors. No one after me is to alter this Treaty; it stands unchangeable for ever.

(Arabic signature.)

We, the Undersigned, do hereby declare that this Treaty was this day read and translated in our presence before the Sultan of Gandu, who stated that it was given with his own hand, and was approved and accepted by him, and was thereupon handed over by him to W. Wallace.

W. WALLACE.
T. M. TEED.
Q. F. GOMES.
T. F. JOSEPH.
(Arabic signature.)

4th July 1894.

I, W. Wallace, for and on behalf of the Company, do hereby approve and accept the above confirmation of previous Treaties and Treaty now made, and hereby affix my hand.

W. WALLACE.

App. 20.—*BRITISH ORDER IN COUNCIL respecting Matabeleland.* 18th July, 1894.*

(Extract.)

WHEREAS the territories of South Africa, situated within the limits of this Order as hereinafter described, are under the protection of Her Majesty the Queen:

And whereas by treaty, grant, usage, sufferance, and other lawful means Her Majesty has power and jurisdiction in the said territories:

Now, therefore, Her Majesty by virtue and in exercise of the powers by the Foreign Jurisdiction Act, 1890, or otherwise in Her Majesty vested, is pleased, by and with the advice of Her Privy Council, to order, and it is hereby ordered, as follows:—

This Order is divided into parts, as follows :—

	Articles.
Part I. Interpretation and application	3— 6
„ II. Administration and Legislation ..	7—25
„ III. Judicial	26—43
„ IV. Land Commission	44—54
„ V. Judicial Notice. Commencement .	55—57

Part I, § 4. The limits of this Order are the parts of South Africa bounded by the Portuguese Possessions, the South African Republic to a point opposite the mouth of the River Shashi, by the River Shashi, and the territories of the Chief Khama of the Bamangwato to the River Zambesi, and by that river to the Portuguese boundary, including an area of 10 miles radius round Fort Tuli, and excluding the area of the district known as the Tati districts, as defined by the Charter (**No. 37**).

* "London Gazette," 27th July, 1894.

App. 21.—*BOUNDARY AGREEMENT between France and the Congo Free State.* 14th August, 1894.

THE undersigned, Gabriel Hanotaux, Minister for Foreign Affairs of the French Republic, &c.; Jacques Haussmann, Director of Political and Commercial Affairs at the Colonial Office, &c.; Joseph Devolder, ex-Minister of Justice and ex-Minister of the Interior and Education of His Majesty the King of the Belgians, Vice-President of the Supreme Council of the Congo Free State, &c.; and Baron Constant Goffinet, &c.; Plenipotentiaries of the French Republic and of the Congo Free State, deputed to prepare an agreement relative to the boundaries of the respective possessions of the two States and to settle the other questions pending between them, have agreed upon the following provisions:—

Boundary between the Congo Free State and French Congo.
Oubanghi, &c.

ART. I.—The frontier between the Congo Free State and the colony of French Congo, after following the thalweg of the Oubanghi up to the confluence of the Mbomou* and of Ouelle [or Welle], shall be constituted as follows:—(1) The thalweg of the Mbomou up to its source. (2) A straight line joining the watershed between the Congo and Nile basins. From this point the frontier of the Free State is constituted by the said watershed up to its intersection with longitude 30° East of Greenwich (27° 40′ E., Paris).

ART. II. *French Right of Police over the Waters of the Mbomou.*

Renunciation by Free State of Occupation or Influence over certain Districts. Watershed of Congo and Nile Basins, &c.

ART. IV. The Free State binds herself to renounce all occupation, and to exercise in the future no political influence west

* The terms "Mbomou" and "Source of the Mbomou," have reference to the indications contained in Junker's map (Gotha, Justus Perthes, 1888).

or north of a line thus determined:—Longitude 30° E. of Greenwich (27° 40' E., Paris), starting from its intersection of the watershed of the Congo and Nile basins, up to the point where it meets the parallel 5° 30', and then along that parallel to the Nile.

Art. V. *Ratifications to be exchanged within Three Months.*

Art. VI. In token of which the Plenipotentiaries have drawn up the present arrangement and affixed their signatures. Given at Paris in duplicate, August 14, 1894.

G. HANOTAUX.
J. HAUSSMANN.
J. DEVOLDER.
Baron GOFFINET.

[Approved by the Government of the French Republic, by Law of 21st December, 1894.]

[Uganda.]

App. 22.—TREATY. Great Britain and Uganda. British Protectorate. 27th August, 1894.

[Approved by Her Majesty's Government, January 4, 1895.]

TREATY between Henry Edward Colvile, a Companion of the Most Honourable Order of the Bath, a Colonel in Her Majesty's army, Her Britannic Majesty's Acting Commissioner for Uganda, for and on behalf of Her Majesty the Queen of Great Britain and Ireland, Empress of India, &c., her heirs and successors, and Mwanga, King of Uganda, for himself, his heirs, and successors. 27th August, 1894.

1. WHEREAS Her Majesty's Government has sanctioned the Agreement between Mwanga, King of Uganda, and Sir Gerald Herbert Portal, K.C.M.G., C.B., Her Britannic Majesty's Commissioner and Consul-General for East Africa, made at Kampala on the 29th day of May, 1893 (page 995);

2. And whereas Her Britannic Majesty has been graciously pleased to bestow on the said Mwanga, King of Uganda, the protection which he requested in that Agreement:

3. I, the said Mwanga, do hereby pledge and bind myself, my heirs, and successors, to the following conditions:—

[Here follow, word for word, the same Articles, 4 to 15, as appear in the Treaty of 29 May, 1893, page 995.]

16. The present Treaty supersedes all other Agreements or Treaties whatsoever made by Mwanga or his predecessors.

17. This Treaty shall come into force from the date of its signature.

In faith whereof we have respectively signed this Treaty, and have thereunto affixed our seals.

Done in duplicate at Kampala this 27th day of August, 1894.

H. E. COLVILE, *Colonel.*
KABAKA, *King.*

Witnesses:
W. T. ANSORGE.
APOLLO, *Katiriko.*
MUGWANYA, *Katiriko.*

App. 23.—*AGREEMENT between Germany and Portugal. Spheres of Influence, East Africa. Kionga, &c. September,* 1894.

It has been publicly stated that an agreement was entered into between Germany and Portugal, in September, 1894, for defining their respective spheres of influence in East Africa, and that both Governments have agreed to recognize as the boundary of their respective possessions the parallel of 10° 40' S. lat. from the coast to the point at which it intersects the river, which thence becomes the common frontier.

This arrangement gives to Germany the mouth of the Rovuma and Kionga Bay, while Tunghi Bay remains to Portugal; but no official copy of this Agreement was obtainable at the time of this volume going to press (February, 1896).

App. 24] BRITISH SOUTH AFRICA COMPANY. [24 Nov., 1894.

[North of the Zambesi.]

App. 24.—*MEMORANDUM of Agreement with South Africa Company respecting British Central Africa, supplementary to the Agreement of February*, 1891 (*p.* 987). *24th November*, 1894.*

Direct Administration by the Company of portion of British Sphere north of the Zambesi.

THE South Africa Company having intimated that it is prepared to undertake the direct administration of the portion of the British sphere north of the Zambesi over which its charter was extended in 1891 (page 987), the arrangement under which the administration was confided to Her Majesty's Commissioner and Consul-General, in consultation with the Company, till the 1st January, 1896, or such earlier date as the Secretary of State might direct, will terminate from the date of the assumption by the Company of direct administration, which shall not be later than the 30th June, 1895.

Administrative posts which have been placed by the Commissioner in the Chartered territory will be transferred, and will thenceforth be under the direct control of the Company. The expenditure of the Commissioner on their account will cease from the date of transfer.

Expense of Police Force.

2. The Company will, in accordance with the existing arrangement, pay into the hands of the Commissioner the annual contribution to the expense of police of 10,000*l*. up to the 1st January, 1896.

Steamers of African Lakes Company.

It will during the year 1895 pay 1,000*l*. in liquidation of the obligation, which it undertook in 1891 (page 989), to afford to the Commissioner the use, free of charge, for administrative purposes, of the steamers on the lake belonging to the African Lakes Company.

* Parl. Paper, Africa, No. 2 (1893).

Outstanding Accounts:—Police Force; Steam Transport on Lake; Expenses of Chartered Territory as distinct from the Protectorate.

3. The outstanding accounts between the Company and the Protectorate will be regulated by the accountants of the Foreign Office with those of the Company, on the basis that the Company is liable only for the annual police contribution of 10,000*l.*, for 5,000*l.* given in 1891 (page 988) for expenses connected with raising and organizing the police force, for expenditure in connection with steam transport on the lake for administrative purposes, and for amounts which can be shown to have been expended for the benefit of, or on account of, the Chartered territory as distinct from the Protectorate.

Sum expended for Operations against Makanjira.

4. It having been explained that Mr. Rhodes voluntarily authorized Her Majesty's Commissioner to spend on his behalf a maximum sum of 10,000*l.* for the operations against Makanjira, it is agreed that the sum actually expended on that account shall be ascertained by the above-mentioned accountants, and that the balance, if any, of the total amount of 10,000*l.*, if the whole amount shall be shown to have been drawn by the Commissioner, shall be repaid by Mr. Rhodes.

Conditional Confirmation of Treaties made by Company in Chartered Territory.

5. The Treaties made on behalf of the Company in the Chartered territory will be sanctioned, on the condition that no provisions in them will be confirmed which may conflict with the prohibition against monopolies contained in the Charter, and with the stipulations of the Act of Berlin (**No. 17**), in so far as they are applicable to the Chartered territory.

Claims based on Concessions.

It is understood that this sanction is without prejudice to certain claims based on Concessions said to have been obtained by Mr. Wiese, should the validity of such Concessions be hereafter established.

[North of the Zambesi.]

Mining Rights in Marimba.

6. The mining rights in the territory described as Marimba in the Commissioner's published map, purchased by the Company from the African Lakes Company, will be confirmed, subject to the terms accepted by the agent of the latter Company.

Mining Rights in Central Angoniland.

In the territory described in the above-mentioned map as Central Angoniland, the mining rights claimed by the Company will be confirmed, subject to the terms accepted by the agent of the African Lakes Company.

These confirmations will not include sanction of administrative powers, monopolies, nor the right to prevent the acquirement of land by settlers.

Company's Claim to Land and Minerals acquired from African Lakes Company.

7. The claim of the Company to land and minerals acquired from the African Lakes Company by purchase in the territory described in the Commissioner's map as North Nyasa,* will be examined when the titles obtained by the latter Company from the native Chiefs are produced for investigation.

German Territory between Lakes Nyasa and Tanganyika to be guarded by Company against aggression.

8. It being necessary that the frontier between Lakes Nyasa and Tanganyika should be watched in order to prevent aggression by the natives on German territory, and the introduction of arms and ammunition in contravention of the prohibition which has been imposed, the Company agrees to take the requisite steps for guarding that frontier.

Customs Arrangements.

9. Customs arrangements between the Protectorate and the

* See Parl. Paper, Africa No. 6 (1894).

Chartered territory which experience may make it desirable to adopt for the purposes of the execution of the Berlin and Brussels Acts (**Nos. 17, 18**), or for fiscal reasons, will be subject to the approval of the Secretary of State.

The Company undertakes to provide Customs posts, and to make suitable arrangements for insuring the proper observance, and preventing abuse, of the stipulations as to free transit in favour of countries adopting the Free Zone system of the Act of Berlin (**No. 17**).

H. PERCY ANDERSON.

(For the British South Africa Company),
C. J. RHODES.

24th November, 1894.

App. 25.—*CONVENTION between Great Britain and the South African Republic, respecting Swaziland. Signed 10th December, 1894.**

WHEREAS Her Majesty the Queen of the United Kingdom of Great Britain and Ireland, and his Honour the State President of the South African Republic, as representing the Government of the said Republic, have agreed that it is expedient that they should enter into a Convention relative to the affairs of Swaziland in substitution of the Conventions of 1890 (vol. 2, page 868) and 1893. (See Note, page 903.)

Now, therefore, Her Majesty the Queen of the United Kingdom of Great Britain and Ireland, and his Honour the State President of the South African Republic, as representing the Government of the said Republic, hereby consent and agree that the following Articles, accepted finally by and between Her Majesty and his Honour shall, when duly signed, sealed, and executed by Her Majesty's High Commissioner for South Africa, on behalf of Her Majesty, and by his Honour the State President of the South African Republic, on behalf of the Government of the said Republic, and when duly ratified by the Volksraad of the South African Republic, constitute and be a Convention by and between Her Majesty the Queen of the United Kingdom of Great Britain and Ireland and the South African Republic.

Conditional Continuance in force of the Convention of 1890.

ART. I. The provisions of the Convention of 1890 shall be continued in full force and effect from and after the date of the signing of this Convention by His Excellency Sir Henry Brougham Loch, Her Majesty's High Commissioner, on behalf of Her Majesty, and his Honour Stephanus Johannes Paulus Kruger, State President of the South African Republic, on behalf of the Government of the South African Republic, until the date of the ratification of this Convention by the Volksraad of the South African Republic; provided that should this Con-

* "Further Correspondence respecting the Affairs of Swaziland," was laid before Parliament in 1895 (C. 7611).

vention not be ratified before or during the next ordinary session of the said Volksraad, the provisions of the Convention of 1890, saving the provisions of Articles 10 and 24 thereof, which shall remain in full force and effect, may at any time thereafter be terminated by one month's notice, given either by Her Majesty's Government, or the Government of the South African Republic, and thereupon, at the expiration of the said month, in accordance with the Convention of 1884 (**No. 179**, vol. 2, page 847) all the provisions relative thereto in the said Convention shall be of full force and effect; and provided further that if at any time before the ratification in manner aforesaid, the assent of the Swazie Queen-Regent and Council to the draft Organic Proclamation already agreed to by Her Majesty's Government and the Government of the South African Republic be duly signified,* the Convention of November, 1893 (**No. 189**), shall, upon the signification of such assent, be and remain of full force and effect, subject to the terms of the said Organic Proclamation, and this Convention shall not thereafter be ratified, but shall be of no force and effect, and the provisions of the Convention of 1890 (vol. 2, page 868) shall no longer be of any force or effect, saving the provisions of Articles 10 and 24 thereof, which shall remain of full force and effect.

Rights of South African Republic over Swaziland secured, subject to certain Conditions.

ART. II. Without the incorporation of Swaziland into the South African Republic, the Government of the South African Republic shall have and be secured in all rights and powers of protection, legislation, jurisdiction, and administration over Swaziland and the inhabitants thereof, subject to the following conditions and provisions, namely :—

1. That the young King Ungwane *alias* Uhili *alias* Ubunu, after he has become of age, according to native law, shall be and remain the Paramount Chief of the Swazies in Swaziland, with the usual powers of such Paramount Chief, in so far as the same are not inconsistent with civilized laws and customs.

* The Organic Proclamation was not assented to by the Swazie Queen-Regent and Council.

2. That the payment by the Government of the South African Republic of monies derived from the collection of the private revenue of the King shall be regularly made in terms of concession or power of attorney, granted in that behalf by Umbandine, and confirmed by the judgment of the chief court.

3. That the management of the internal affairs of the natives shall be in accordance with their own laws and customs, including the laws and customs of inheritance and succession, and that the native laws and customs shall be administrated by the native chiefs entitled to administer the same in such manner as they are in accordance with the native law and custom at present administering, in so far as the said laws and customs are not inconsistent with civilized laws and customs, or with any law in force in Swaziland made pursuant to this Convention, and the natives are guaranteed in their continued use and occupation of land now in their possession, and of all grazing or agricultural rights to which they are at present entitled; provided that no law made hereafter in Swaziland shall be in conflict with the guarantees given to the Swazies in this Convention.

4. That in the administration and government of the country by the Government of the South African Republic, no hut tax or other tax shall be imposed upon the natives higher than the corresponding tax to which such of the Swazie people as are living within the borders of the Republic may be subject. In no case, however, shall such taxes be able to be imposed until after the expiration of three years from the date of the ratification of this Convention.

Appointment of an Administrator of Swaziland by South African Republic.

ART. III. The Government of the South African Republic agrees to appoint an officer who shall administer Swaziland in terms of this Convention.

Powers and Jurisdiction of the Chief Court.

ART. IV. The Government of the South African Republic agrees that the chief court heretofore established shall continue to exercise and possess all the powers and jurisdiction hitherto exercised or possessed by it; the said court shall also have such powers and jurisdiction as may be conferred upon it, in accordance with Article II of this Convention, subject to the conditions of the said Article, with full power, to decree against all persons, execution of every order, judgment, decree, or sentence made by it in the exercise of its jurisdiction.

Confirmation of Existing Swaziland Laws, Ordinances, &c.

ART. V. The laws ordinances, proclamations, and regulations at present in force in Swaziland shall continue to be of full force and effect therein until altered, amended, or repealed in accordance with the terms of this Convention; and the power and jurisdiction heretofore exercised or possessed by Landdrost Courts and justices of the peace shall continue to be exercised and possessed by such courts and such justices of the peace respectively, unless and until other provision be made in accordance with the terms of this Convention.

Government Officers appointed under Convention of 1890.

ART. VI. All Government officers appointed under and by virtue of the Convention of 1890 (vol. 2, page 868), shall continue to hold and administer the offices to which they have been appointed, and shall be secured in the emoluments and fees of office at present enjoyed by them, until the date of the ratification of this Convention, or until other provision be made in that behalf by Her Majesty's Government or the Government of the South African Republic, and thereupon all such appointments shall cease and determine; provided that on or after the date of ratification aforesaid the said officials or any of them may be reappointed to the said offices or any of them, in accordance with the terms of this Convention.

Confirmation of all Rights, &c., of British Subjects and their Property in Swaziland.

ART. VII. All British subjects residing in Swaziland, or having in Swaziland any property, grant, privilege, or concession, or any right, title to, or interest in, any property, grant, privileges, or concession, shall be secured in the future enjoyment of all their rights and privileges of whatsoever nature or kind in like manner as burghers of the South African Republic, but shall obey the Government and conform to the laws established for Swaziland.

Political Privileges of every White Male resident in Swaziland on 20th April, 1893, subject to certain Conditions.

ART. VIII. Every white male who shall have been a *bonâ fide* resident in Swaziland (even if temporarily absent from Swaziland) on the 20th April, 1893, shall become and be entitled to all the political privileges of a full burgher of the South African Republic as though he had been born in that Republic: provided, however—

(*a.*) That every white male shall make application in writing to an officer to be appointed at Bremersdorp, in Swaziland, by the Government of the said Republic, to have his name enrolled upon a list of persons so entitled, and upon satisfactory proof by a true and solemn declaration of his *bonâ fide* residence in Swaziland on the aforesaid day, such declaration to be made within six months from the date of public notification of the appointment of such officer as aforesaid, such officer shall be bound to enrol his name on such list, and such list shall be the list of burghers of the South African Republic so admitted under this head of this article to the privileges aforesaid.

(*b.*) That every white son of any person admitted to the privileges of a burgher under the preceding paragraph of this article, which son shall have been a minor on the aforesaid date, shall be entitled to the like political privileges which he would have had if his father had been a natural-born burgher of that Republic and he himself had been born therein, provided that the right under this section shall be claimed by such minor from the Government of the South African Republic by notice in writing within 12 months from his attaining his majority.

(*c.*) That every person admitted as a burgher shall, while

resident in Swazieland, be entitled to register his vote at any election when and where a burgher resident in some convenient district of the South African Republic adjoining Swaziland would be entitled to vote, such district to be determined by the Government of the South African Republic, and if thereafter he shall come to reside in any district of the South African Republic such person shall there be entitled to register his vote.

Use of Dutch and English Languages in Courts of Law in Swaziland.

ART. IX. The equal rights of the Dutch and English languages in all Courts of Swaziland shall be maintained. This provision shall be in force so long as the administration of Swaziland by the Government of the South African Republic continues under the provisions of this Convention.

Customs Duties on Articles imported into Swaziland.

ART. X. The Customs duties shall not be higher in respect of any article imported into Swaziland than the duty thereon according to the tariff at present in force in the South African Republic, or the tariff at present in force in the South African Customs Union, whichever is now the higher. This provision shall be in force so long as the administration of Swaziland by the Government of the South African Republic continues under the provision of this Convention. Every exclusive right or privilege of or belonging to any individual or individuals, corporation or company, with regard to imposition of or exemption from customs duties on goods shall be liable to expropriation by the administering authority; provided that no such individual or individuals, corporation or company, shall be deprived of or interfered with in the enjoyment of any such exclusive rights or privileges as have been confirmed by the Chief Court prior to the 8th November, 1893, without due compensation being awarded. The amount of such compensation shall be assessed by means of arbitration in case of difference. Each party interested shall appoint an arbitrator, and the said arbitrators shall, before proceeding with the arbitration, appoint an umpire; should the said arbitrators be unable to agree upon an umpire such umpire shall, upon application of either party, after notice to the other, be appointed by the Chief Court; the decision of the majority of the persons so appointed shall, in case of difference, be final.

Prohibition against sale of Intoxicating Liquors to Swazies in Swaziland.

ART. XI. The Government of the South African Republic agrees to prohibit the sale or supply of intoxicating liquor to Swazie natives in Swaziland.

Railways.

ART. XII. No railway beyond the eastern boundary of Swaziland shall be constructed by the Government of the South African Republic save under the provisions of a further contemplated Convention between Her Majesty the Queen and the South African Republic, or with the consent of Her Majesty's Government.

Articles 10 and 24 of Convention of 1890.

ART. XIII. Articles 10 and 24 of the Convention of 1890 (vol. 2, page 868) are here again set forth for convenience of reference.

"Article 10. The Government of the South African Republic withdraws all claims to extend the territory of the Republic, or to enter into treaties with any natives or native tribes to the north or north-west of the existing boundary of the Republic, and undertakes to aid and support by its favouring influence the establishment of order and government in those territories by the British South Africa Company within the limits of power and territory set forth in the Charter granted by Her Majesty to the said Company."

Inclusion of the Little Free State within the South African Republic.

"Article 24. Her Majesty's Government consent to an alteration of the boundary of the South African Republic on the east so as to include the territory known as the Little Free State within the territory of the South African Republic."

Diplomatic Representation in favour of Swazies, &c. British Rights Reserved.

ART. XIV. Her Majesty's Government reserves the power of exercising diplomatic representation in favour of Swazie natives or British subjects in case any provision of this Convention shall not be fairly and faithfully observed.

Right Reserved to appoint a British Consular Officer in Swaziland.

ART. XV. Her Majesty's Government reserves the right to appoint a British Consular Officer to reside in Swaziland.

Signed and Sealed on the border of Natal and the South African Republic, near Charlestown and Volksrust, this 10th day of December, 1894.

HENRY B. LOCH, *High Commissioner.*

Signed and Sealed on the border of Natal and the South African Republic, near Charlestown and Volksrust, this 10th day of December, 1894.

S. J. P. KRUGER, *State President of the South African Republic.*

Dr. W. J. LEYDS, *Staats Secretaris, Z.A.R.*

Ratification by Volksraad.

[This Convention was ratified by the Volksraad of the South African Republic on the 13th February, 1895.]

App. 26.—*DECLARATION between Great Britain and Portugal, agreeing to refer to Arbitration the Boundary in dispute under Art. II of the Treaty of 11th June, 1891. Signed at London, 7th January, 1895.*

(Translation.)

ON the 11th June, 1891, a Treaty was signed between Her Majesty the Queen of the United Kingdom of Great Britain and Ireland, Empress of India, and His Most Faithful Majesty the King of Portugal and the Algarves, which Treaty settled the question of the boundaries of their Possessions and Spheres of Influence in Eastern and Central Africa (No. 150).

Art. 2 of this Treaty contains the demarcation of the boundary to the south of the Zambesi, that is to say, from the point on the bank of this river opposite the mouth of the Aroangoa or Loangwa, as far as the point where the boundary of Swaziland intersects the River Maputo.

Differences having arisen with regard to the meaning of certain phrases in the said Article, the two Governments have decided to have recourse to the arbitration of His Excellency Mr. Paul Honoré Vigliani, formerly First President of the "Cour de Cassation," Senator and Minister of State of the kingdom of Italy.

They do not, however, propose that the whole of the above-mentioned line should be submitted to the arbitration.

The boundary to the south of the Zambesi may be considered as divided into three sections:—

1. From the Zambesi as far as 18° 30′ S. lat.
2. From 18° 30′ S. lat. to a point where the Rivers Sabi and Lunde, or Lunte, meet.
3. From this point to the River Maputo.

It is not considered necessary to submit to arbitration the line defined in Sections 1 and 3; the differences only concern the 2nd section.

The negotiations took place in London. The text of the Treaty was drawn up in English, and initialled by the Marquis of Salisbury, then Minister for Foreign Affairs, and by M. de Soveral, Portuguese Minister. The Treaty having been compared with the copy initialled in London, was signed at Lisbon by Count Valbom, Portuguese Minister for Foreign Affairs, and by Sir George Petre, Her Britannic Majesty's Minister at Lisbon.

That portion of the Article which deals with the second section of the boundary is drawn up in the following terms:—

"From there" (that is to say, from the intersection of the 33rd degree of longitude east of Greenwich and of the parallel 18° 30' of S. lat.) "it follows in a southerly direction the upper portion of the eastern slope of the Manica plateau as far as the middle of the principal channel of the River Sabi, and follows that channel as far as the point where it meets the Lunte.

"It is understood that in tracing the boundary along the slope of the plateau no portion of territory to the west of the meridian 32° 30' of long. E. of Greenwich shall be included in the Portuguese sphere, nor any portion of territory to the east of the meridian 33° of long. E. of Greenwich in the British sphere. Nevertheless, should the case arise, the line shall be drawn so as to leave Mutassa in the British sphere and Massi-Kessi in the Portuguese sphere."

The following are the terms, in English and Portuguese:—

. . . . "Thence it follows the upper part of the eastern slope of the Manica plateau southwards to the centre of the main channel of the Sabi, follows that channel to its confluence with the Lunte, whence it strikes direct to the northeastern point of the frontier of the South African Republic, and follows the eastern frontier of the Republic, and the frontier of Swaziland, to the River Maputo.

. . . "D'ahi accompanha a crista da vertente oriental do planalto de Manica na sua direcção sul até á linha media do leito principal do Save, seguindo por elle até á sua confluencia com o Lunde, d'onde corta direito ao extremo nordeste da fronteira da Republica Sul Africana, continuando pelas fronteiras orientaes d'esta Republica, e da Swazilandia até ao Rio Maputo.

[Boundary. Manica Plateau.]

"It is understood that in tracing the frontier along the slope of the plateau, no territory west of longitude 32° 30' east of Greenwich shall be comprised in the Portuguese sphere, and no territory east of longitude 33° east of Greenwich shall be comprised in the British sphere. The line shall, however, if necessary, be deflected so as to leave Mutassa in the British sphere, and Massi-Kessi in the Portuguese sphere."

"Fica entendido ao traçar a fronteira ao longo da crista do planalto, nenhum territorio a oeste do meridiano de 32° 30' de longitude leste de Greenwich será comprehendido na esphera Portugueza, e que nenhum territorio a leste do meridiano de 33° de longitude leste de Greenwich ficará comprehendido na esphera Britannica. Esta linha soffrera comtudo, sendo necessario, a inflexão bastante para que Mutassa fique na esphera Britannica e Macequece na esphera Portugueza."

In the month of June, 1892, the Commissioners of the two Governments endeavoured to trace the boundary line according to the above-mentioned stipulations, but a difference having arisen between them, the settlement was referred to their Governments. Direct negotiations between the Ministry for Foreign Affairs of Lisbon and the Foreign Office have taken place; but all prospect of arriving at an understanding having appeared impossible, the two Governments have decided to have recourse to arbitration.

These diplomatic negotiations and the technical labours of the Commissioners have left the question of demarcation in the following position:

1. As regards the territory comprised between the parallel 18° 30' and a point situated at a distance of a few miles to the south of the Chimanimani Pass, each Commissioner has proposed a boundary line, and each Government has adopted the line proposed by his Commissioner; whence a difference of opinions has arisen which they have not yet found means of reconciling.

2. As regards the territory comprised between a point

[Boundary. Manica Plateau.]

situated at a distance of a few miles to the south of the Chimanimani Pass and the parallel 20° 42' 17" of S. lat., the British Commissioner and a Delegate of the Portuguese Commissioner, as far as he was authorized, have agreed upon a boundary line, the examination of which by the two Governments has re-remained unfinished.

3. As regards the territory which extends from the parallel 20° 42' 17" of S. lat. as far as the point where the rivers Sabi and Lunte meet, no project of demarcation has been discussed between the two Governments.

In these circumstances, the two Governments have agreed to request the Arbitrator to take into consideration the documents, the reports of the negotiations, and the results of the technical labours, to weigh the arguments of the two Governments based upon their respective opinions, and to decide on the line which shall separate the Portuguese sphere of influence from that of Great Britain from the parallel 18° 30' to the point of confluence of the Lunte and Sabi.

In faith of which the undersigned duly authorized by their respective Governments have signed the present Declaration, to which they have affixed the Seals of their Arms.

Done in London, on the 7th January, 1895.

(L.S.) KIMBERLEY.
(L.S.) LUIZ DE SOVERAL.

App. 27.—*TREATY for the Cession of the Congo Free State to Belgium. Signed at Brussels, 9th January,* 1895.

[Submitted to Belgian Chamber of Deputies for approval, 13th February, 1895, but not yet approved. 1st February, 1896.]

THE King-Sovereign of the Congo having made known, in his letter of the 5th August, 1889, to the Belgian Minister of Finance, that if it suited Belgium to enter into, before the allotted time, closer relations with her Congo Possessions, His Majesty would not hesitate to place them at her disposal; and the two High Contracting Parties having agreed to carry out this Cession at once;

The following Treaty has been concluded between the State of Belgium, represented by Count de Merode Westerloo, Minister for Foreign Affairs, M. de Burlet, Minister of the Interior, and of Public Instruction, and M. de Smet de Naeyer, Minister of Finance, acting subject to the approval of the Legislature,

And the Congo Free State, represented by M. E. Van Eetvelde, Secretary of State of the said Free State:

ART. I. His Majesty the King-Sovereign declares that he cedes from the present time to Belgium the Sovereignty of the territories constituting the Congo Free State with all the rights and obligations attached thereto, and the State of Belgium declares that she accepts this cession.

ART. II. The Cession comprises all real and moveable property (mobilier) of the Free State, and especially:

1. The ownership of all the land belonging to her public or private domain, subject to the obligations and duties indicated in Annex A to the present Convention.

2. The Shares and Founders' Shares (Parts de Fondateurs) which have been allotted to her in the formation of the Railway Company, as well as all shares or shares of interest (Parts d'Interêts) which have been allotted to her in the arrangements mentioned in Annex A.

3. All houses, buildings, settlements, plantations, and other

property whatsoever, established or acquired by the Government of the Free State, moveable objects of every kind, and cattle which she may possess, her ships and boats with their gear, as well as her military stores.

4. The ivory, india-rubber, and other African produce which are actually the property of the Free State, as well as the provisions and other goods belonging to her.

ART. III. On the other hand, the Cession comprises all the liabilities and all the financial engagements of the Free State, as set forth in detail in Annex B.

ART. IV. The date on which Belgium shall assume the exercise of her right of sovereignty over the territories mentioned in Article I shall be determined by a Royal Decree.*

The monies received and the expenses incurred by the Free State, on and after the 1st January, 1895, shall belong to Belgium.

In faith of which, the respective Plenipotentiaries have signed the present Treaty and have affixed thereto their seals.

Done in duplicate at Brussels, on the 9th January, 1895.

(L.S.) COUNT DE MERODE DE WESTERLOO.
(L.S.) J. DE BURLET.
(L.S.) P. DE SMET DE NAEYER.
(L.S.) EDM. VAN EETVELDE.

[Here follow Annexes A and B.]

* Not yet issued (February 1, 1896).

App. 28] CONGO. [11 Jan., 1895.

[Neutrality. Congo State.]

App. 28.—*DECLARATION of the Neutrality of the Congo Free State.* 11th *January*, 1895.*

(Translation.)

THE rule of neutrality which formed the subject of the Declaration notified on the 1st August, 1885 (**No. 42**), to the Signatory Powers of the General Act of the Berlin Conference (**No. 17**) shall henceforth apply to the territory of the State delimitated as follows, in consequence of the Protocol of the 29th April, 1887 (**No. 51**), and of the Arrangement of the 14th August, 1894, concluded with the French Republic (vol. 2, page 1002), of the Conventions concluded on the 25th May, 1891 (**No. 59**), and of the Declarations signed on the 24th March, 1894 (page 986), with the Government of His Most Faithful Majesty, and of the arrangement concluded on the 12th May, 1894 (**No. 990**), with the British Government:—

To the North.

A straight line, 950 metres long, starting from a point on the beach of the Atlantic Ocean 300 metres to the north of the principal house of the Dutch factory at Lunga, a point, the latitude of which is 5° 47′ 14″ 31 S., and joining, in a south-easterly direction, the mouth of the rivulet Lunga, which flows into the lagoon of the same name;

The course of the rivulet Lunga, as far as Mallongo Pool—the villages of Congo, N'Conde, Jema, &c., remaining to the Congo Free State—those of Cabo-Lombo, M'Venho, Jabe, Ganzy, Taly, Spita-Gagandjime, N'Goio, M'To, Fortalisa, Sokki, &c., to Portugal;

The courses of the Rivers Venzo and Lulofe, as far as the source of the latter on the slope of Mount Nime-Tchiama, the geographical bearings of this source being 5° 44′ 19″ 60 S. lat., 12° 17′ 25″ 28 long. E. of Greenwich;

* Laid before the Belgian Chamber of Representatives, 13th February, 1895, with other documents, when submitting for its approval the Treaty of Cession concluded 9th January, 1895, between Belgium and the Independent State of the Congo (p. 79), but which Treaty has not yet been formally approved (1st February, 1896).

1043

[Neutrality. Congo State.]

The parallel of this source, as far as its intersection with the meridian of the confluence of the Luculla with the river called by some N'Zenze, and by others Culla-Calla, the bearings of this confluence being 5° 10' 49" 30 S. lat., 12° 32' 06" 60 long. E. of Greenwich.

The meridian, thus determined, until it meets the River Luculla;

The course of the Luculla as far as its confluence with the Chiloango (Loango-Luce);

The River Chiloango, from the mouth of the Luculla as far as its northernmost source;

The watershed of the Niadi-Kuilou and of the Congo, as far as the meridian of Manyanga;

A line to be determined which following as far as possible a natural division of the land, shall terminate between the station of Manyanga and the cataract of Ntombo-Mataka, to a point situated on the navigable portion of the river;*

The Congo as far as Stanley Pool;

The median line of Stanley Pool;

The Congo as far as its confluence with the Oubanghi;

The thalweg of the Oubanghi as far as the confluence of the M'Bomou with the Ouellé;

The thalweg of the M'Bomou up to its source;

A straight line meeting the watershed of the Congo and Nile Basins.

To the North-East.

The watershed of the Nile and Congo up to its intersection by the meridian 30° E. of Greenwich (27° 40' Paris);

The extension of this watershed until its second intersection by the aforesaid meridian 30° E. of Greenwich.

To the East.

The 30th degree of long. E. of Greenwich up to 1° 20' of S. lat.;

A straight line drawn from the intersection of the 30th

* It was stated in a foot-note that this line had been pratially determined.

degree of E. long. with the parallel 1° 20' S. lat. to the northern extremity of Lake Tanganika;

The median line of Lake Tanganika;

A line running straight to the extremity of Cape Akalunga, on Lake Tanganika, situated on the northernmost point of Cameron Bay, by about 8° 15' S. lat., to the right bank of the River Luapula, at the point where the river leaves Lake Moëro;*

From this point, a line drawn straight to the mouth of the River Luapula in Lake Moëro; this line, moreover, deviates towards the south of the lake so as to leave the island of Kilwa to Great Britain;

The thalweg of the Luapula up to the point where this river leaves Lake Bangweolo;

The meridian in a southerly direction, passing through this point, to the watershed of the Congo and Zambesi.

To the South.

The watershed of the Congo and Zambesi to the source of the affluent of the Kassaï which rises in Lake Dilolo;

The course of this affluent from its source to its mouth;

The thalweg of the Kassaï as far as the parallel 7° 17' S. lat.;

The parallel 7° 17' S. lat. to its intersection by the thalweg of the Chikapa;

The thalweg of the River Chikapa to its intersection by the parallel 6° 55' S. lat.;

The parallel 6° 55' S. lat. to its intersection by the thalweg of the Lovua;

The thalweg of the Lovua to its intersection by the 7th degree of S. lat.;

The 7th degree of S. lat. to its intersection by the thalweg of the Loangué;

The thalweg of the Loangué to its confluence with the Kangulungu or Kama Bomba;

The thalweg of the Kangulungu to its intersection by the

* See Agreement, Great Britain and Congo State, 12th May, 1894. Art. I (b), p. 1009.

[Neutrality. Congo State.]

parallel of the confluence of the Kwilu and the Luita (about 7° 34′ S. lat.);

This parallel as far as the confluence of the Kwilu and the Luita;

The thalweg of the Luita from the junction of its waters with the Kwilu to the 8th degree of S. lat.;

The 8th degree of S. lat. as far as its intersection by the thalweg of the Kwengo;

The thalweg of the Kwengo to its intersection by the parallel 7°.55′ S. lat.;

The parallel 7° 55′ of S. lat. as far as the thalweg of the Lucaïa;

The thalweg of the Lucaïa as far as the 8th degree of S. lat.;

The 8th degree of S. lat. as far as the thalweg of the Kamanguna, the river through whose waters the River Lué enters the N'Kombo;

The thalweg of the Kamanguna, and of the N'Komba, to its junction with the Uövo;

The thalweg of the Uövo as far as its mouth in the Wamba;

The thalweg of the Wamba from the mouth of the Uövo to its intersection by the parallel of the point of junction of the Komba and Lola (8′ west of the Wamba, and at about 8° 5′ 40″ S. lat.);

The thalweg of the channel along which the waters of the Lola flow to its intersection by the thalweg of the Tungila;

The thalweg of the Tungila to its mouth in the Kwango (about 80° 7′ 40″ S. lat.);

The thalweg of the Kwango until it meets the parallel passing through the residence at Nokki;

The parallel passing through the residence at Nokki (5° 52′ 10″ 14 S. lat., 13° 28′ 25″ 25 long. E. of Greenwich) from the Kwango to a point determined on this parallel, 2000 metres to the east of the left bank of the Congo;

A straight line joining this point with the point of intersection of the left bank of the Congo by the parallel which runs 100 metres to the north of the principal house of the factory of Domingos de Souza, at Nokki;

[Neutrality. Congo State.]

This parallel as far as its intersection by the median line of the navigation channel generally used by ships drawing a large amount of water;

This median line, as far as the mouth of the Congo River, which line actually leaves to the right, notably, amongst others, the river islands named Bulambemba, Mateba, and Princes Isles, which are comprised between this line and the right bank of the river, and to the left notably, amongst others, the river islands known by the names of Bulicoco and Sacran Abaca Islands, which are comprised between this line and the left bank of the river.

To the West.

The Atlantic Ocean between the point where the aforesaid median line terminates at the sea and the point situated 300 metres to the north of the principal house of the Dutch factory at Lunga.

Brussels, 28th December, 1894.

App. 29.—*AGREEMENT between Great Britain and France, fixing the Boundary between the British and French Possessions to the North and East of Sierra Leone. Signed (in the English and French languages) at Paris, 21st January,* 1895.*

English Version.

THE Special Commissioners nominated by the Governments of Great Britain and France, in accordance with Article V of the Agreement of August 10th, 1889 (**No. 110**), having failed to trace a line of demarcation between the territories of the two Powers to the north and east of Sierra Leone, in conformity with the general provisions of Article II of the said Agreement, of its Annex I and of its Annex II (Sierra Leone), and with the indications of the Agreement of June 26th, 1891 (**No. 114**), the under-signed Plenipotentiaries charged, in execution of the declarations exchanged at London, on August 5th, 1890 (**No. 113**), between Her Britannic Majesty's Government and the Government of the French Republic, to proceed to delimit the respective spheres of interest of the two countries south and west of the Middle and Upper Niger, have agreed to fix the line of demarcation between the above-mentioned territories on the following conditions:—

ART. I. The boundary starts from a point on the Atlantic coast north-west of the village of Kiragba, where a circle of 500 metres radius, described from the centre of the village, cuts high-water mark.

From this point it proceeds, in a north-easterly direction, parallel to the road leading from Kiragba to Robenia (Roubani), which passes by or near the English villages of Fungala, Robant, Mengeti, Mandimo, Momotimenia, and Kongobutia, at an even distance of 500 metres from the centre of the track, as far as a point half-way between the village of Kongobutia (English) and the village of Digipali (French). From this point it turns to the south-east, and, cutting the road at right angles, reaches a point 500 metres on the south-eastern side,

---

* Par. Pap. Treaty Series No. 5 (1895). A map defining this boundary was laid before Parliament with this paper.

[Boundary North and East of Sierra Leone.]

and proceeds parallel to the road, at an even distance of 500 metres, measured as before from the centre of the track, till it reaches a point to the south of the village of Digipali, whence it is drawn directly to the watershed formed by a ridge which, commencing south of the destroyed village of Passinodia, distinctly marks the line of separation between the basin of the Mellakori (Mellacorée) River and that of the Great Skarcies or Kolenté River.

The frontier follows this watershed line, leaving to Great Britain the villages of Bogolo (N'Bogoli), Musaliya, Lukoiya (Malaguia), Mufuri (Maforé), Tarnenai (Tanéné), Modina (Madina), Oblenia, Oboto, Ballimir, Massini, and Gambiadi; and to France, the villages of Robenia (Roubani), N'Tunga (N'Tugon), Daragli (Daragoué), Kunia, Tombaiya, Heremakuno (Erimakono), Fransiga (Fonsiga) Talansa, Tanganne (Tagani), and Maodea, as far as the point nearest to the source of the Little Mola River; from this point it follows a straight line to the above-metioned source, follows the course of the Little Mola to its junction with the Mola, and then the thalweg of the Mola to its junction with the Great Skarcies or Kolenté.

From this point the frontier follows the right bank of the Great Skarcies (Kolenté) as far as a point situated 500 metres south of the spot where the road leading from Wulia (Ouelia) to Wossu (Ouossou), *viâ* Lucenia, touches the right bank. From this point it crosses the river and follows a line drawn to the south of the above-mentioned road at an even distance of 500 metres, measured from the centre of the track until it meets a straight line connecting the two points mentioned below, namely:—

1. A point on the Kora, 500 metres above the bend of the river, which is situated about 2,500 metres north of the village of Lusenia, or about 5 kilom. up the Kora River, measured along the bank, from its point of junction with the Great Skarcies (Kolentó).

2. A gap in the north-western face of the chain of hills lying in the eastern part of Talla, situated about two English miles (3,200 metres) south of the village of Duyunia (Donia).

From this point of intersection it follows the above-mentioned line eastwards to the centre of the above-mentioned gap, from whence it is drawn straight to a point on the River Kita, situated above and at a distance of 1,500 metres, as the crow flies, from the centre of the village of Lakhata. It then follows the thalweg of the Kita River as far as the confluence of that stream with the Lolo.

From this point of junction it coincides with a line drawn straight to a point on the Little Skarcies or Kaba River, four English miles (6,400 metres) south of the 10th parallel of north latitude; and it then follows the thalweg of the Little Skarcies as far as the said parallel, which then forms the boundary as far as its intersection with the watershed ("ligne de partage des eaux"), separating the basin of the Niger on the one hand, from the basins of the Little Skarcies and other rivers, falling westward to the Atlantic Ocean on the other hand.

Finally the frontier follows the aforesaid watershed southeastward, leaving Kalieri to Great Britain and Herimakuna (Erimakono) to France, until its intersection with the parallel of latitude passing through Tembi-kunda (Tembikounda), that is to say, the source of the Tembiko or Niger.

ART. II. The boundary defined in this Agreement is marked on the map which is annexed hereto.

ART. III. This Agreement is regarded by the two Governments as completing and interpreting Article II of the Agreement of 10th August, 1889 (**No. 110**), Annex 1 of the said Agreement, Annex 2 of the said Agreement (heading Sierra Leone), and the Agreement of the 26th June, 1891 (**No. 114**).

Done at Paris, the 21st January, 1895.

(L.S.) E. C. H. PHIPPS.
(L.S.) J. A. CROWE.
(L.S.) GEORGES BENOIT.
(L.S.) J. HAUSSMANN.

ANNEX.

Maps.

Although the delineation of the line of demarcation on the map annexed to the present Agreement is believed to be generally accurate, it shall not be considered as an absolutely

[Boundary North and East of Sierra Leone.]

correct representation of that line until it has been confirmed by future surveys.

It is therefore agreed that in the event of Commissioners or local Delegates of the two countries being hereafter appointed to delimit the whole or any portion of the frontier on the ground, they shall be guided by the description of the frontier as set forth in the Agreement. They shall at the same time be permitted to modify the said line of demarcation for the purpose of delineating its direction with greater accuracy, and also to rectify the position of the watersheds, roads, or rivers, as well as that of any of the towns or villages indicated on the map above referred to.

It is, however, understood that any alterations or corrections proposed by common consent of the aforesaid Commissioners or Delegates shall be submitted for the approval of their respective Governments.

(Translation.)

Tembi-Kunda.

(1). *M. Hanotaux to the Marquis of Dufferin.*

M. l'Ambassadeur, Paris, 22nd January, 1895.

DURING the course of the recent discussions relative to the delimitation of the French and British possessions to the north and east of Sierra Leone, the Commissioners of the two countries have been led to examine the situation resulting from the Arrangement concluded on the 8th December, 1892 (**No. 164**, page 783), between the Government of the French Republic and the Government of the Republic of Liberia, in so far as the eastern frontier of the British Colony of Sierra Leone is concerned, and they have agreed upon the following declaration :—

" According to the Arrangement concluded on the 8th December, 1892 (**No. 164**, page 783), between the Government of the French Republic and the Government of the Republic of Liberia, the frontier line between the French possessions and the Republic of Liberia is fixed by the parallel of Tembi-kunda until it meets, at the 13th degree of longitude west of Paris, the Anglo-French frontier of Sierra Leone.

[Boundary North and East of Sierra Leone.]

" The delimitation of the Anglo-French frontier of Sierra Leone, therefore, terminates at the parallel of Tembi-kunda.

" At the same time it is necessary to recall that, in virtue of the Notes exchanged on the 2nd December, 1891, and the 4th March, 1892, between M. Ribot and Mr. Egerton, the 13th degree of longitude west of Paris was in any case to constitute the limit of the French Soudanese possessions and the British Colony of Sierra Leone up to the point of the intersection of that meridian with the Anglo-Liberian frontier.

" It was under such circumstances that the French Government ceded to the Liberian Government certain territories forming part of the French Soudan, situated to the south of the parallel of Tembi-kunda, and to the east of the 13th degree of longitude west of Paris.

" It is accordingly understood, that from the point of intersection of the watershed separating the basin of the Niger on the one hand from the basins of the rivers flowing westwards to the Atlantic Ocean on the other hand, with the parallel of latitude passing through Tembi-kunda, the frontier of the Colony of Sierra Leone is formed by the said parallel as far as the 13th degree of longitude west of Paris, and then by that meridian until it meets the Anglo-Liberian frontier."

I have the honour to inform your Excellency that the Government of the French Republic is disposed to approve the terms of this Declaration, and I shall be obliged if you will be so good as to inform me whether the Government of Her Britannic Majesty also assent to it.

Accept, &c.,

G. HANOTAUX.

Tembi-Kunda.

(2.) *The Marquis of Dufferin to M. Hanotaux.*

M. le Ministre,   Paris, 22nd January, 1895.

I HAVE the honour to acknowledge the receipt of your Excellency's note of the 22nd instant, in which you observe that, during the course of the recent discussions relative to the delimitation of the British and French territories to the north and east of Sierra Leone, the Commissioners of the two countries had been led to examine the situation resulting from the Arrangement concluded on the 8th December, 1892 (**No. 164**), between the Government of the French Republic and the Government of the Republic of Liberia in so far as the eastern frontier of the Colony of Sierra Leone is concerned; and that the Commissioners had agreed upon the following Declaration :—

" According to the Arrangement concluded on the 8th December, 1892 (**No. 164**), between the Government of the French Republic and the Government of the Republic of Liberia, the frontier line between the French possessions and the Republic of Liberia is fixed by the parallel of Tembi-kunda until it meets, at the 13th degree of longitude west of Paris, the Anglo-French frontier of Sierra Leone.

" The delimitation of the Anglo-French frontier of Sierra Leone, therefore, terminates at the parallel of Tembi-kunda.

" At the same time it is necessary to recall that, in virtue of the Notes exchanged on the 2nd December, 1891, and the 4th March, 1892, between Mr. Egerton and M. Ribot, the 13th degree of longitude west of Paris was in any case to constitute the limit of the French Soudanese possessions and the British Colony of Sierra Leone up to the point of the intersection of that meridian with the Anglo-Liberian frontier.

" It was under such circumstances that the French Government ceded to the Liberian Government certain territories forming part of the French Soudan situated to the south of the parallel of Tembi-kunda, and to the east of the 13th degree of longitude west of Paris.

" It is accordingly understood that, from the point of intersection of the watershed separating the basin of the Niger on the one hand from the basins of the rivers flowing westwards to the Atlantic Ocean on the other hand, with the parallel of latitude passing through Tembi-kunda, the frontier of the Colony of Sierra Leone is formed by the said parallel as far as the 13th degree of longitude west of Paris, and then by that meridian until it meets the Anglo-Liberian frontier."

I have the honour, under instructions from Her Britannic

Majesty's Government, to inform your Excellency that Her Majesty's Government is disposed to approve the terms of the declaration as above embodied.

I have, &c.,

DUFFERIN AND AVA.

Use of Open Roads by Traders and Travellers.

(3.)—*M. Hanotaux to the Marquis of Dufferin.*

(Translation.)

M. l'Ambassadeur,   *Paris, 22nd January,* 1895.

DURING the course of the recent discussions relative to the delimitation of the French and British possessions to the north and east of Sierra Leone, the Commissioners of the two countries arrived at an understanding as to the principle of the arrangements intended to regulate the commercial relations between the British Colony of Sierra Leone and the neighbouring French possessions. It was at the same time understood that the conditions of this understanding should form the subject of an exchange of notes immediately after the signature of the Agreement.

In consequence, I have the honour to inform your Excellency that the Government of the French Republic is disposed to give its assent to the following stipulations:—

1. In the territories dependent on the Colony of Sierra Leone, on the one hand, and in those dependent upon the Colonies of French Guinea (including Fouta Djallon) and of the French Soudan, on the other hand, the traders and travellers belonging to the two countries shall be treated upon a footing of perfect equality in so far as the use of roads and other means of land communication are concerned.

2. The roads crossing the frontier indicated by the Agreement of the 21st January, 1895 (Vol. i, page 82), between the British Colony of Sierra Leone and the neighbouring French Colonies shall on both sides be open to commerce on payment of such duties and taxes as may be established.

3. The two Governments reciprocally engage not to estab-

[Boundary North and East of Sierra Leone.]

lish on the land frontier defined by the Agreement of the 21st January, 1895, between their respective Colonies, any duties, either import or export, higher than those which shall be levied on the maritime frontier either of the Colony of Sierra Leone or of the Colony of French Guinea.

The duties on exports shall not in any case exceed 7 per cent. *ad valorem*, calculated according to the official Tables of Valuation of each Colony.

4. Posts at which the duties or taxes on imports and exports shall be paid shall be established at certain fixed points on the frontier in order that caravans may not be diverted from the roads which they might desire to follow in order to pass from the Colony of Sierra Leone into the neighbouring French Colonies, or *vice versâ*.

I shall be obliged to your Excellency if you will be so good as to inform me whether the Government of Her Britannic Majesty are on their part disposed to give their consent to the arrangement in question.

Accept, &c.,

HANOTAUX.

Use of Open Roads by Traders and Travellers.

(4).—*The Marquis of Dufferin to M. Hanotaux.*

M. le Ministre,   *Paris, 22nd January,* 1895.

I HAVE the honour to acknowledge the receipt of your Excellency's note of the 22nd instant, in which you observe that during the course of the recent discussions relative to the delimitation of the British and French posessions to the north and east of Sierra Leone, the Commissioners of the two countries had arrived at an understanding as to the principle of the arrangements intended to regulate the commercial relations between the British Colony of Sierra Leone and the neighbouring French possessions. Your Excellency points out that it was at the same time understood that the conditions of this understanding should form the subject of an exchange of notes immediately after the signature of the Agreement.

In consequence, your Excellency does me the honour of intimating to me that the Government of the Republic is disposed to give its assent to the following stipulations:—

1. In the territories dependent on the Colony of Sierra Leone, on the one hand, and in those dependent upon the Colonies of French Guinea (including Fouta Djallon) and of the French Soudan, on the other hand, the traders and travellers belonging to the two countries shall be treated upon a footing of perfect equality in so far as the use of roads and other means of land communication are concerned.

2. The roads crossing the frontier indicated by the Agreement of the 21st January, 1895 (page 82), between the British Colony of Sierra Leone and the neighbouring French Colonies shall on both sides be open to commerce on payment of such duties and taxes as may be established.

3. The two Governments reciprocally engage not to establish on the land frontier defined by the Agreement of the 21st January, 1895 (Vol. i, page 82), between their respective Colonies, any duties, either import or export, higher than those which shall be levied on the maritime frontier either of the Colony of Sierra Leone or of the Colony of French Guinea.

The duties on exports shall not in any case exceed 7 per cent. *ad valorem*, calculated according to the official Tables of Valuation of each Colony.

4. Posts at which the duties or taxes on imports and exports shall be paid shall be established at certain fixed points on the frontier in order that caravans may not be diverted from the roads which they might desire to follow in order to pass from the Colony of Sierra Leone into the neighbouring French Colonies, or *vice versâ*.

I am instructed by Her Majesty's Government to express their acceptance of the arrangement above recorded, which they have no doubt will prove beneficial to the trading and commercial interests of the two countries.

I have, &c.,

DUFFERIN AND AVA.

[Boundary North and East of Sierra Leone.]

Continued use of Right Bank of Great Skarcies River by Riverain Inhabitants.

(5.)—*The Marquis of Dufferin to M. Hanotaux.*

M. le Ministre,   Paris 22nd January, 1895.

DURING the course of the recent negotiations relative to the delimitation of the British and French territories and possessions situated to the north and east of Sierra Leone, the Commissioners named by the two Powers were led to examine the situation created to the riverain inhabitants of a certain portion of the Great Skarcies by the execution of the Agreement of the 10th August, 1889 (**No. 110**).

Although by Article I of the Agreement of the 21st January, 1895, the British frontier follows the right bank of the Great Skarcies from a point on the right bank, 500 metres south of the road leading from Wulia to Wossu, *viâ* Lusenia, to the point where that river is joined by the Little Mola, Her Majesty's Government is, nevertheless, disposed to permit the riverain inhabitants dwelling on the right bank within the above-mentioned limits to continue to use the river to the same extent as heretofore.

It is, however, understood that the inhabitants of these villages will be subject to such Laws or Ordinances as may from time to time be promulgated by the authorities of the Colony of Sierra Leone with a view to regulating the navigation of the river or in connection with the control of its waters, due notice of the same being given by the Governor of Sierra Leone to the Governor of French Guinea.

I have, &c.,

DUFFERIN AND AVA.

(6.) *M. Hanotaux to the Marquis of Dufferin.*
(Translation.)

M. l'Ambassadeur,   Paris, 4th February, 1895.

I HAVE received the letter which your Excellency did me the honour to address to me on the 22nd January last, on the subject of the exchange of views which has taken place between the Commissioners of the two countries in the course of the recent negotiations relative to the delimitation of the French and British possessions to the north and east of Sierra Leone, respecting the situation created to the riverain inhabitants of a certain portion of the Great Skarcies by the execution of the Agreement of the 10th August, 1889 (**No. 110**).

Your Excellency informs me that although by the terms of Article I of the Agreement of the 21st January, 1895, the British frontier follows the right bank of the Great Skarcies from a point situated on the right bank, 500 metres south of the road leading from Wulia to Wossu, *viâ* Lusenia, to the point where the Little Mola flows into that river, Her Majesty's Government is, nevertheless, disposed to permit the inhabitants dwelling in the villages on the right bank within the above-mentioned limits to continue to use the river under the same conditions as heretofore.

It is, however, understood that the inhabitants of these villages will be subject to such Laws and Ordinances as may be promulgated by the authorities of the Colony of Sierra Leone, with a view of regulating the navigation of the river, or the police of its waters, after due notice of the same shall have been given by the Governor of Sierra Leone to the Governor of French Guinea.

I hasten to thank your Excellency for this communication, which I have not failed to make known to the Minister for the Colonies.

I have, &c.,

G. HANOTAUX.

[French Right of Pre-emption. Congo State.]

App. 30.—*ARRANGEMENT entered into between the Belgian Government and France, respecting the French right of pre-emption over the Territories of the Congo State. Signed at Paris, 5th February, 1895.**

(Translation.)

Arrangement regulating the Right of Preference of France over the Territories of the Congo State, 5th February, 1895.

WHEREAS by virtue of notes exchanged $\frac{23rd}{24th}$ April, 1884 (**No. 46**), between M. Strauch, President of the International Association of the Congo, and M. J. Ferry, President of the Council and Minister for Foreign Affairs of the French Republic, a right of preference (*préférence*) was secured to France in the event of the association being some day induced to dispose of its possessions; and whereas this right of preference was maintained when the Independent State of the Congo replaced the International Association (**No. 50**);

Whereas, in view of the transfer to Belgium of the possessions of the Independent State of the Congo, by virtue of the Treaty of Cession of 9th January, 1895 (page 1041), the Belgian Government will take over the obligation (*se trouvera substitué à l'obligation*) contracted in this respect by the Government of the said State;

The undersigned have agreed upon the following dispositions that shall regulate henceforth the French right of preference with respect to the Belgian Colony of the Congo.

ART. I. The Belgian Government recognizes a right of preference to France over its Congolese Possessions in case of alienation of the same for a consideration, either in whole or in part.

Every exchange of Congolese territories with a Foreign Power, every concession, every leasing of the said territories, in whole or in part, to a Foreign Power, or to a foreign Company invested with rights of sovereignty, shall equally be made subject to the French right of preference, and shall consequently form the object of a preliminary negotiation between

* Not yet approved (1st February, 1896).

[French Right of Pre-emption. Congo State.]

the Belgian Government and the Government of the French Republic.

ART. II. The Belgian Government declares that it will never cede gratuitously the whole or any part of the said possessions.

ART. III. The dispositions provided for by the above Articles apply to the whole of the territories of the Belgian Congo.

In testimony whereof the undersigned have drawn up the present arrangement, to which they have affixed their seals.

Done in duplicate, at Paris, 5th February, 1895.

      (L.S.)   BON D'ANETHAN.
      (L.S.)   G. HANOTAUX.

[This Arrangement was submitted to the Belgian Chamber of Deputies for approval, on the 13th February, 1895; but it has not yet been approved (1st February, 1896).]

App. 31.—*DECLARATION exchanged between Belgium and France, relative to the Limits of their respective Possessions in Stanley-Pool. Paris, 5th February,* 1895.*

(Translation.)

Declaration exchanged between the Belgian Government and the Government of the French Republic, 5th February, 1895.

Stanley Pool.

THE Belgian Government and the Government of the French Republic agree to adopt as limits of their respective possessions in Stanley-Pool:

Island of Bamou.

The median line of Stanley-Pool up to the point of contact of this line with the island of Bamou, the southern shore of this island up to its eastern extremity, then the median line of Stanley-Pool.

The island of Bamou, the waters and islets inclosed between the island of Bamou and the northern shore of Stanley-Pool shall belong to France; the waters and islets inclosed between the island of Bamou and the southern shore of Stanley-Pool shall belong to Belgium.

Military establishments shall not be created in the island of Bamou.

In testimony whereof the undersigned have drawn up the present Declaration, to which they have affixed their seals.

Done in duplicate, at Paris, 5th February, 1895.

(L.S.) BON D'ANETHAN.
(L.S.) G. HANOTAUX.

---

* The Treaty between the Congo State and Belgium has not yet been ratified (1st February, 1896).

App. 32.—*CONVENTION between Spain and Morocco Indemnity, &c.* 24th February, 1895.

Supplementary Convention to that with Morocco of the 5th March, 1894 (**No. 902**). *Signed in Madrid, 24th February, 1895.*

(Translation.)

In the Name of Almighty God!

[Here follow the names of the Plenipotentiaries.]

ART. I.—*Chastisement of the Riffians guilty of Outrages in October—November, 1893.*

ART. II.—*Demarcation of Polygonal Line. Boundaries of Neutral Zones near Fortress of Melilla, postponed for another year.*

ART. III.—*Establishment and Maintenance of Moorish Troops in neighbourhood of Melilla.*

ART. IV.—*Balance of Indemnity of 1,000,000 dollars to be paid by Morocco to Spain within 80 days, or an Annual Interest at Rate of 6 per cent. to be payable whilst Capital is in arrear.*

ART. V.—*Remaining 14 Instalments of Indemnity due to Spanish Government may be made in one payment in Gold during next six months.*

ART. VI.—*Interest at rate of 6 per cent. per Annum to be paid should any unavoidable delay occur. Right of Spain to "intervene" in Moorish Custom Houses, in event of Indemnity not being punctually paid.*

Confirmation of previous Treaties.

ART. VII. In so far as they are not modified by the present Convention, all the stipulations specified in previous Treaties between Spain and Morocco, and principally that of the 5th March, 1894 (vol. 2, page 902), and those referring to the fortress and territory of Melhilla, shall remain in force.

ART. VIII. The present Convention shall be ratified, and the ratifications shall be exchanged in Tangier within the term of 40 days from the date of signing the same.

Wherefore the respective Plenipotentiaries sign it in duplicate, and seal it with their seals, in Madrid, on the 24th February, 1895, of the Christian Era, corresponding to the 29th Shabaan, 1312.

  (L.S.) ALEJANDRO GROIZARD.
  (L.S.) SIDI-HADJ-EL-KERIM BRISCHA.

The present Convention has been duly ratified, and the ratifications exchanged in Tangier on the 4th April, 1895.

App. 33.—*AGREEMENT between the British and Moorish Governments, respecting the purchase by Morocco of the property of the North-West Africa Company in Terfaya (Cape Juby). Signed, 13th March, 1895.*

(Translation.)

AGREEMENT as concluded between the two persons who are going to sign at the end of this document, and they are—the Vizier, the honoured, the worthy Cid Hamad-ben-Moosa-ben-Hamad, and the gentleman the Minister, Mr. Satow; and they have agreed to the six following clauses below, concerning the Government (Moorish) buying, from the English Company called the North-West African, the buildings, &c., in the place that is known by the name Terfaya, that is, in the country of the tribe of Tekna.

Lands between Wad Draa and Cape Bojador belong to Morocco.

1st Clause.—If this Government buy the buildings, &c., in the place above named from the above-named Company, no one will have any claim to the lands that are between Wad Draa and Cape Bojador, and which are called Terfaya above named, and all the lands behind it, because all this belongs to the territory of Morocco.

Such Lands not to be given to any other Power.

2nd Clause.—It is agreed that this Government shall give its word to the English Government that they will not give any part of the above-named lands to anyone whatsoever without the concurrence of the English Government.

All Property of Company to be handed over to Morocco for 50,000l.

3rd Clause.—If this Government buy the buildings in the place above named from the Company above named, the whole of the property shall belong to them, namely, the buildings with their stones and wood, that are on the land or out at sea (*i.e.*, the reef), and the whole of the property that is inclosed in the walls of the buildings, whether on the land or at sea, including cannons and any other property, and no one shall be able to lay claim of any kind whatsoever to the above proper-

* Page 809.

[North-West Africa Company. Cape Juby.]

ties or lands; and the price this Government is to pay for all this to the above-named Company is put down at 50,000*l.*, half at the signing of this document, the other half when the Government receives over into their hands the above-named lands from the Company above mentioned.

Places belonging to late Company to remain open to Trade.
Customs' Duties.

4th Clause.—If the Moorish Government take over the place named from the Company named by buying it, it shall remain open for buying and selling, and the Customs' duties for exports and imports shall be the same as at other ports on the coast.

Moorish Government not to be compelled to build Houses, &c., for Merchants.

5th Clause.—If the Moorish Government take over the place named from the Company named by buying it, the Moorish Government shall not build from the money of the Treasury any houses for the merchants to live in, or stores for their merchandize, and shall not supply boats to land or ship cargo until such time as it may please the Sultan to do so.

Right of Merchants to build Houses, &c., at their own expense.

6th Clause.—If any merchants wish to bring merchandize to the place named, and take a letter from the Minister of their nation, this Government shall allot to them a piece of ground at a rental to build suitable stores or dwelling-houses, at the merchants' own expense, for 20 years, and at the end of 20 years, the said allotments, with the buildings thereon, shall become the property of the Moorish Government.

Agreement of Sultan of Morocco to above Clauses.

(After compliments.)

I have shown the six clauses written above to the Sultan. God give him the victory about the Agreement between us concerning these six clauses about the buying for the Government of our Lord the buildings of the place named above. The Emperor—God help him—agreed to them all, and allowed them all. Also he grants his consent to the buying of the buildings for his Government—God prosper them—from the

Company named above for 50,000*l.*, half of it at once, and the other half when the Government receive over the place named, which shall be within six months, counting from the 1st Shawal next to the end of Rebia I next, and the Sultan—God bless his soul—has ordered me to write the above. And also the Government perhaps will get ready some people belonging to them to go the place above named at once, before they receive it over, and when they send them they will let you know, so that you can give them a letter from you to the Englishmen there, so that they will receive them.

      HAMAD-BEN-MOOSA-BEN-HAMAD.
16 *Ramadan*, 1312 (*March* 13, 1895).

Agreement of British Envoy to above Clauses.

Supplementary Clause.

To the worthy, honoured, and wise Vizier, Cid Hamad-ben-Moosa-ben-Hamad.

I agree to the six clauses written above, and I also agree to the Company above named selling the buildings at the place above named to the Government of the Sultan—may God bless him—for a sum of 50,000*l.*, the Government to pay half at once, the other half within six months, counting from the 1st Shawal next (28th March) to the end of Rebia I next (19th September), and the transfer of the place above named to the Moorish Government by the Company above named shall take place whenever the Moorish Government pays down the remaining half, namely, 25,000*l.*, to the above-named Company.

In token whereof I hereto append my signature, this 13th day of March, 1895, being duly authorized thereto by Her Britannic Majesty's Government.

If the Moorish Government desire to send any officials to reside at Cape Juby there is no objection, but before doing so they must let me know, that I may write a letter to the Englishmen in charge there to receive them.

      ERNEST SATOW, *Her Britannic Majesty's Envoy Extraordinary and Minister Plenipotentiary.*

App. 34]   GREAT BRITAIN (AMATONGALAND).   [11 June, 1895.

[Tongaland.]

App. 34.—*PROCLAMATION. British Sovereignty over Territories of certain Native Chiefs in Zululand (Amatongaland, Maputaland, or Tembeland).* 23rd April, 1895.

PROCLAMATION, in the name of Her Most Gracious Majesty, Victoria, by the Grace of God, of the United Kingdom of Great Britain and Ireland, Queen, Defender of the Faith, Empress of India, &c.

WHEREAS it is expedient that the territories bounded on the south and east by the Pongola River, on the North by the Maputa or Usutu River, and on the West by Swaziland and the South African Republic, being the territories of the Native Chiefs Umbegeza, Mdhlaleni, Sambane or Zambaan, and of other Native Chiefs therein residing, should be added to the dominions of Her Majesty Queen Victoria :

And whereas Her Majesty has been pleased to authorize m to take the necessary steps for giving effect to her pleasure in the matter:

Now, therefore, I, Walter Francis Hely-Hutchinson, Knight Commander of the Most Distinguished Order of Saint Michael and Saint George, Governor of the Territory of Zululand, do hereby, by command of Her Most Gracious Majesty Victoria, by the Grace of God, of the United Kingdom of Great Britain and Ireland, Queen, Defender of the Faith, Empress of India, &c., conveyed to me through Her Principal Secretary of State for the Colonies, proclaim and declare to all men that the full Sovereignty of the territories bounded on the south and east by the Pongola River, on the north by the Maputa or Usutu River, and on the west by Swaziland and the South African Republic, is vested in Her Most Gracious Majesty, Queen Victoria, her heirs and successors for ever.

God Save the Queen !

Given under my hand and seal at Pietermaritzburg, Natal, this 23rd day of April, 1895.

By command of His Excellency the Governor of Zululand,

H. M. TABERER,

Secretary for Zululand.

App. 35.—*BRITISH NOTIFICATION. British Protectorate over part of Amatongaland (Maputaland or Tembeland). Natal, 11th June, 1895.*

GOVERNMENT NOTICE.

His Excellency the Governor of Zululand directs it to be notified, for general information, that, on the 30th May, 1895, at Ngwanasi's Kraal, Mr. C. R. Saunders, the Resident Magistrate of the Eshowe District of Zululand, the officer selected by the Governor of Zululand to carry out the instructions of Her Majesty's Government to that effect, formally declared a British Protectorate over the territory variously known as Amatongaland, Maputaland, or Tembeland, bounded on the north by the southern boundary of the Portuguese territory, viz., by a line following the parallel of the confluence of the Pongolo River with the Usutu or Maputa River to the Indian Ocean; on the east by the Indian Ocean; on the west by the eastern boundary of the territories added to Her Majesty's dominions by the Zululand Proclamation of the 23rd April, 1895 (page 1067), viz., by the Pongolo River; and on the south by the territory of Zululand.

By his Excellency's command,

W. E. PEACHEY,
Acting Secretary for Zululand.

Government House, Pietermaritzburg,
Natal, 11th June, 1895.

App. 36.—*NOTIFICATION. British Protectorate over Territories in East Africa, late in possession of the British East Africa Company. Foreign Office,* 15*th June,* 1895.

Foreign Office, June 15, 1895.*

It is hereby notified for public information that the territories in East Africa under the influence of Great Britain, lying between the Protectorate of Uganda and the coast, and between the River Juba and the northern frontier of the German sphere, not being already under British protection, are placed under the protectorate of Her Britannic Majesty.

Ceremony of the formal Transfer of the Territory administered by the Imperial British East Africa Company to Her Britannic Majesty's Government.

Speech of Sir Lloyd Mathews, Wazir of the Sultan of Zanzibar at Mombasa, in the "baraza," held on the 1st July, 1895.

"Governor, Sheikhs, Elders, and all people of the country under our Lord, Seyyid Hamed-bin-Thwain,

"I have come here to-day by order of our Lord, Seyyid Hamed-bin-Thwain, to inform you all that the Company have retired from the administration of his territory, and that the great English Government will succeed it, and Mr. Hardinge, the Consul-General at Zanzibar will be the head of the new Administration, and will issue all orders in the territory under the sovereignty of His Highness. And all affairs connected with the faith of Islam will be conducted to the honour and benefit of religion, and all ancient customs will be allowed to continue, and his wish is that everything should be done in accordance with justice and law."

---

* "London Gazette," 18th June, 1895.

15 June, 1895.] BRITISH EAST AFRICA. [App. 36
[British Protectorate.]

SPEECH of Mr. A. H. Hardinge, H.M.'s Consul-General at Zanzibar, read in Arabic and translated into Swahili.

"In accordance with what you have heard from the letter of his Highness Seyyid Hamed-bin-Thwain, your Sultan, and with what his Wazir has just told you, I announce to you that from to-day:—

Administration of Sultan of Zanzibar's Dominions on the Mainland by Officers under Control of the British Consul-General at Zanzibar.

"I take over, in the name of the great Government, the administration of this country, and of all the countries inland as far as Kikuyu, and of the whole coast from Wanga to Kismayu.

Part belonging to Zanzibar to be under Sultan's Sovereignty, but under British Administration.

"You know that a part of these territories belongs to your Lord the Seyyid; this part is and remains under his sovereignty but, I shall be its Administrator and Governor, according to the Agreement which has been made between himself and my exalted Government.* And the regions of the interior will be placed under officers whom I shall appoint in obedience to the commands of the great Government, of which you will be informed at a future date.

Mahommedan Law and Religion. Religious Liberty to all.

"And with respect to what the Wazir of the Sultan has told you about religion, let it be known to you that it will be protected and respected by the new Administration, and that all mosques and religious festivals, and Cadis and Ulema will receive all honour at our hands. The Mahommedan religion will remain the public and established creed in the Sultan's territory, and all cases and lawsuits between natives will continue to be decided according to the 'Shoira,' but although

* See Agreement, 31st August, 1889. Vol. 2, p. 760.

the Mahommedan is and remains the State religion, we intend that there shall be the fullest liberty for all others, and that all their adherents, whether they be Christians, or Parsees, or Hindoos, shall freely worship God according to their respective rites.

Projected Railway.

"We are resolved to rule these territories in accordance with justice and equity, and to strive to promote the happiness of their inhabitants, and I trust that the new Administration, and especially the railway, which the High Government has now decided shall be made, will conduce to the welfare and prosperity of the country.

Confirmation of Company Officers in their positions.

"Lastly, I confirm the present Administrator at Mombasa, and all Walis, Cadis. Akidas, and other officers of the former Company in their present positions pending further orders, and I enjoin upon you all to continue to obey them."

[The standard of the Sultan, as territorial Sovereign, was then saluted with 21 guns by Her Majesty's ship "Phœbe," and at the same moment the Imperial British East Africa Company's Flag was lowered from the top of the Government building on which it formerly flew, the Consular Union Jack being hoisted in its place. This concluded the ceremony of the transfer.]

[Frontier between the Baraka and the Red Sea.]

App. 37.—*AGREEMENT between the Egyptian and Italian Governments for regulating the Dependence of the semi-nomadic Tribes, and for defining their respective Frontiers between the Baraka and the Red Sea. 25th June,* 1895.

WITH a view to establishing in a permanent manner the dependence of the semi-nomadic tribes which exist on the frontier between the Baraka and the Red Sea, and to determining precisely the frontier separating the Italian and Egyptian territory in this region,

His Excellency Ferik Sir Herbert Kitchener Pasha, Sirdar of the Egyptian army; and

His Excellency Cavaliere Oreste Baratieri, Lieutenant-General, Governor of the Colony of Eritrea;

Authorized by their respective Governments, have agreed to the following Articles :—

Boundary Line.

ARTICLE 1. In the district between the Red Sea and the Baraka, the frontier line between Egypt and Eritrea, shall follow a line which, starting from Ras Kassar, joins the principal branch of the Karora, about 2 kiloms. from the coast, and follows the course of the Karora up to a point marked "Karora" on the map. The frontier then follows the watershed between the torrents Aïet and Merib on the north, and the torrents Falkat and Sela on the south, up to a point on the plateau of Hagar-Nush, to be fixed by the delineators; and from the point so fixed shall proceed to join the Baraka at a point which is also left to the delineators to establish, following a clearly determined natural line. From the Baraka the line of frontier goes straight to the intersection of the 17th parallel north, with the 37th meridian east of Greenwich.

Semi-nomadic Tribes on Frontier.

ART. 2. The semi-nomadic tribes on the frontier known as the Hazerandawa (Ad-Azeri), Felunda (Aflenda), Beit Maleh, and Roshaida, together with the sections of the Beni-Amer, at present acknowledging the authority of Sheikh Idris Hamed, are recognized as dependent on the Egyptian Government; and the Beni-Amer acknowledging Sheikhs Ali Hussein and Mahmud Sherif, as well as the Hababs, are recognized as depending on the Government of Eritrea.

Rights of Pasturage and Cultivation in respective Territories.

ART. 3. The two Governments bind themselves to concede reciprocally, reserving to themselves the power of imposing a moderate tax in payment, rights of pasturage, and cultivation in their respective territories to such alien tribes as apply for the concession through their respective Governments. Such concession shall be limited only by the requirements of public safety and by the needs of other local tribes dependent on the Government giving the concession.

Opposition to Immigration of rebellious Tribes into respective Territories.

ART. 4. The two Governments bind themselves reciprocally to oppose, as far as is practicable, the settlement in their respective territories of tribes who may cross the frontier in consequence of rebellion or defection of their Chiefs.

Disarmament of Frontier Tribes.

In the interests of public tranquillity, the two Governments further bind themselves to take into consideration, according to the circumstances of each case, the expediency of total or partial disarmament of the tribes on the frontier, due allowance being made for the requirements of their defence.

Rebellious Tribes on Frontier.

Each of the two Governments, however, reserves to itself full liberty to decide upon the measures which it shall take in its own territory, both as regards the means of opposing the immigration of rebellious tribes, and in respect to the time,

[Frontiers between the Baraka and the Red Sea.]

method, and extent of the disarmament to which this Article refers.

HERBERT KITCHENER,
Sirdar, Egyptian Army.

Cairo, 25th June, 1895.

Subsequently signed by
General BARATIERI,
Italian Agent and Consul-General.

App. 38.—*EXCHANGE OF NOTES between the British and Portuguese Governments, defining the Frontiers of their respective Possessions in the neighbourhood of Tongaland. September—October,* 1895.

(1.) *Sir H. MacDonell to Senhor de Soveral.*

M. le Ministre, *Lisbon,* 24*th September,* 1895.

DURING the course of the recent discussions with regard to the declaration of a British Protectorate over Tongaland, it was agreed that the adoption of the new frontier between the British and Portuguese possessions in that neighbourhood should be recorded by an exchange of Notes.

In consequence, I have the honour to inform your Excellency that Her Majesty's Government agree that the line described in Article III of the Anglo-Portuguese Treaty of the 11th June, 1891 (**No. 150**), shall be the frontier between the territories of the two Powers, that is to say: Her Majesty's Government recognize as belonging to Portugal territory as far south as a line following the parallel of the confluence of the River Pongolo with the River Maputo to the sea coast.

It is agreed that the above line of demarcation shall be subject to rectification by agreement between the two Powers in accordance with local requirements.

I shall be glad if your Excellency will be so good as to inform me whether the Portuguese Government are, on their part, disposed to give their adhesion to the line in question, and to agree to the date of the 15th October next for addressing a simultaneous Notification to the Parties to the Act of Berlin (**No. 17**) to the effect that the new frontier has been definitely accepted by the two Powers.

I avail, &c.,
H. G. MACDONELL.

(2.) *Senhor de Soveral to Sir H. MacDonell.*
(Translation.)

Foreign Department, Lisbon,
Your Excellency, 5th October, 1895.

As it was agreed between your Excellency and my predecessor that the adoption of the new frontier between the Portuguese and British Possessions in Amatongaland should be recorded by an Exchange of Notes, I have the honour to inform your Excellency, in reply to your Note of the 24th September, that His Majesty's Government agree that the line described in Article III of the Luso-British Treaty of the 11th June, 1891 (**No. 150**), shall be the frontier between the territories of the two Powers, that is to say: Her Britannic Majesty's Government recognize as belonging to Portugal the territory as far south as a line following the parallel of the confluence of the Rivers Maputo and Pongolo to the sea.

It is understood that the above-mentioned line of demarcation shall be subject to rectification by agreement between the two Powers in accordance with local requirements.

I have further to inform your Excellency that His Majesty's Government are on their part willing to give their adhesion to the aforesaid line, and to assent to the fixing of the 15th October as the date of the simultaneous Notification to the Signatory Powers of the Act of Berlin (**No. 17**) as regards the definitive acceptance of the new frontier by the two Powers.

I avail, &c.,
LUIZ DE SOVERAL.

NOTIFICATION OF ABOVE TO TREATY POWERS.

(3.) *The Marquess of Salisbury to Her Majesty's Representatives at Courts of the Signatories of the Berlin Act.*

My Lord,
Sir, Foreign Office, 15th October, 1895.

I HAVE to request your Excellency to notify to the Government to which you are accredited, under Article XXXIV of the General Act of the Conference of Berlin (**No. 17**), that the districts on the coast of the African Continent hereinafter described have been formally placed under the Protectorate of Her Britannic Majesty, viz. :—

The territory known as Amatongaland, lying between the British Colony of Zululand, the Portuguese possessions, and the Indian Ocean.

I am, &c.,

SALISBURY.

App. 39.—*BRITISH NOTIFICATION. Non-recognition by Her Majesty's Government of certain concessions granted by Queen Regent of Amatongaland. London, 4th November, 1895.*

Downing Street, 4th November, 1895.

THE attention of Her Majesty's Secretary of State for the Colonies having been drawn to Notices in the London Gazette of the 29th of October, 1895, concerning two Concessions alleged to have been granted by the Queen Regent of Amatongaland on the 11th of May, 1888:

It is hereby notified by Mr. Secretary Chamberlain, for the information of all whom it may concern, that the alleged Concessions are not recognized as valid by Her Majesty's Government, and that the Amatongaland Exploration Company (Limited), which claims to be the holder of these Concessions, was so informed at a date prior to that of the above-mentioned Notices.*

* "London Gazette," 5th November, 1895.

CHRONOLOGICAL LIST.

CHRONOLOGICAL LIST.

				No.	Page
1778.	1 Mar.	Treaty......	Portugal and Spain. Cession. Fernando Po and Annabon to Spain......	181	882
	24 Oct.	Act........	Portugal. Cession. Ditto........	181	884
1783.	3 Sept.	Treaty......	Great Britain and France. Portendic, &c.........................	103	539
1788.	22 Aug.	Declaration.	Sierra Leone. Cessions to Great Britain........................	100	484
1806.	12 Jan.	Capitulation.	Cape of Good Hope to the English..	89	341
	—	Deed......	Purchase. St. Mary Island by Great Britain......................	90	365
1807.	10 July	Treaty......	Great Britain and Sierra Leone. Cession. Bance Island to Great Britain.	100	484
1814.	30 May	Treaty... .	Great Britain and France. Cession. Isle of France, Rodriques, and Les Sechelles to Great Britain........	104	540
	13 Aug.	Convention.	Great Britain and Netherlands. Cession. Cape of Good Hope to Great Britain........................	137	672
1817.	28 July	Convention.	Great Britain and Portugal. Portuguese Limits. East and West Coasts of Africa......................	138	683
1818.	6 July	Treaty......	Great Britain and Bago. Cession. Isles de Los to Great Britain......	100	485
1819.	25 May	Convention.	Great Britain and Timmanees. Cession. Banana Islands to Great Britain.......................	100	486
1820.	21 July	Convention.	Great Britain and Timmanees. Do...	100	487
	20 Oct.	Convention.	Great Britain and Bananas. Do....	100	487
1821.	7 May	Act of Parl..	Abolition. Africa Company	100	488
	5 June	Treaty......	Great Britain and North Bulloms. Cession. Tombo Island to Great Britain.........................	100	488
1823.	14 April	Deed......	Cession. Lemain Island to Great Britain........................	90	365
1824.	2 Aug.	Convention.	Great Britain and North Bulloms. Cession. Bance, Tasso, Tombo Islands, &c., to Great Britain	100	489
1825.	24 Sept.	Convention.	Great Britain and Sherbro. Cessions to Great Britain	100	491
	3 Oct.	Proclamation.	Sherbro and Ya Comba. Territories annexed to Sierra Leone	100	493
	12 Dec.	Convention.	Great Britain and Barra. Cession. Bacea Loco to Great Britain	100	494
	30 Dec.	Agreement..	Great Britain and Mandingo. Cession. Island of Matacong to Great Britain........................	100	496

CHRONOLOGICAL LIST.

				No.	Page
1826.	18 April	Treaty.....	Great Britain and Soombia Soosoos, &c. Cessions to Great Britain. Matacong, &c.....................	100	495
	15 June	Treaty.....	Barra. Cession. Gambia River, &c., to Great Britain.................	90	367
	19 June	Add. Art...	Barra. Boundaries. French Factory, Albreda........................	90	369
1827.	8—10 Mar.	Treaty.....	Great Britain and Kafu-Bulloms. Cessions to Great Britain.........	100	496
	29 May	Treaty.....	Brekama. British Sovereignty. Post opposite Island of Kakaya	90	370
	4 June	Treaty.....	Cession. St. Mary Island, &c......	90	370
	24 June	Agreement.	Great Britain and Biafra. Cession. Island of Bulama to Great Britain.	100	498
1829.	13 April	Treaty.....	Wooli. Cession. Fattatenda to Great Britain........................	90	371
1830.	8 Aug.	Treaty.....	France and Tunis. Coral Fishery. Island of Tabarque	190	906
1831.	27 April	Treaty.....	Great Britain and Ashantee, &c......	91	388
1832.	5 Jan.	Convention.	Barra. Cession to Great Britain. Right bank of River Gambia	90	372
	$\frac{3}{4}$ Oct.	Treaty.....	Ditto, ditto	190	906
1833.	21 Sept.	Treaty.....	Zanzibar and United States. Consular Jurisdiction	207	965
1840.	13 July	Convention.	Combo. Cession to Great Britain...	90	373
	19 Aug.	Deed.......	Tajourah. Sale. Mussa Island to Great Britain	178	832
	27 Aug.	Deed.......	Tajourah. Sale. Island of Bab to Great Britain	178	832
	3 Sept.	Treaty.....	Zeila. Cession. Island of Aubad to Great Britain, &c................	178	832
1841.	13 Feb.	Firmans....	Turkey. Boundaries of Egypt.....	67	259
	23 April	Treaty.....	Cartabar. Cession. Territory to Great Britain........................	90	374
	16 Nov.	Treaty.....	Great Britain and Shoa............	1	2
1843.	12 May	Proclamation.	Great Britain. Natal.............	93	434
	5 Oct.	Treaty.....	Great Britain and Zoolah. St. Lucia Bay. Boundaries, &c...........	93	434
	30 Nov.	Award.....	Prussia. British Claims against France. Portendic	105	541
	Oct.-Dec.	Treaty.....	Great Britain and Basutos. Allegiance.	87	329
1844.	31 May	Letters Pat..	Great Britain. Annexation. Natal. Cape Colony	93	435
	21 Aug.	Proclamation.	Great Britain. Boundaries. District. Natal	93	435
	25 Aug.	Treaty.....	Spain and Morocco. Limits. Ceuta.	184	891
	10 Sept.	Convention.	France and Morocco. Limits......	170	802
	7 Oct.	Treaty.....	Spain and Morocco. Limits. Ceuta.	184	891
	17 Nov.	Treaty.....	France and Madagascar. Consular Jurisdiction, &c.	196	927
	1830-1844	Notes	On French Occupation of Algiers...	75	287
1845.	18 Mar.	Treaty.....	France and Morocco. Limits......	171	803

CHRONOLOGICAL LIST.

Year	Date	Type	Description	No.	Page
1845.	6 May	Treaty	Spain and Morocco. Limits	185	893
1847.	8 Mar.	Treaty	Great Britain and Dahomey. Flags.	65	249
	29 Nov.	Treaty	Great Britain and Loceo Marsamma. Cessions. Sierra Leone River	100	499
	29 Nov.	Treaty	Great Britain and Kafu-Bulloms. Ditto, ditto	100	500
1848.	3 Feb.	Proclamation.	Great Britain. Sovereignty over Orange Free State	174	814
1849.	4—7 July	Treaty	Great Britain and Bompey and Plantain Island. Boundary	100	501
	2 Nov.	Treaty	Great Britain and Abyssinia. Friendship, &c.	2	2
1850.	30 Mar.	Oath	Akropong. Fealty to British Crown.	91	388
	11 Apr.	Treaty	Great Britain and Amapondas. Cession. Boundary. Natal	93	435
	13 July	Order in C.	Great Britain. Jurisdiction. Vicinity, Sierra Leone	100	501
	17 Aug.	Convention.	Great Britain and Denmark. Cessions. Gold Coast	66	256
	18 Nov.	Convention.	Barra. Cession to Great Britain. Territory near Albreda	90	374
	26 Dec.	Convention.	Combo. Territory ceded to Great Britain	90	378
1851.	21 Mar.	Treaty	Great Britain and Naloes. Boundaries.	100	502
	23 Mar.	Letters Pat.	Great Britain. Orange River Territory.	174	81
	2 Aug.	Treaty	Great Britain and Fouricaria. Boundaries	100	502
	23 Dec.	Treaty	Great Britain and Small Scarcies. Boundaries	100	502
	26 Dec.	Treaty	Great Britain and Kambia. Great Scarcies. Boundaries	100	502
1852.	26 Dec., 1851 / 17 Jan., 1852	Treaty	Great Britain and Maebatees (Great Scarcies). Boundaries	100	503
	17 Jan.	Convention.	Great Britain and Transvaal Boers Recognition. Independence	179	830
	29 Jan.	Treaty	Great Britain and Wonkafong. Boundaries	100	503
	1 Mar.	Agreement.	Great Britain and Lagos. Land. Church Missionary Society	92	405
	26 Aug.	Treaty	Great Britain and Kaffu-Bulloms. Cessions. Sierra Leone River	100	503
1853.	24 May	Convention.	Combo. Cessions to Great Britain.	90	380
1854.	30 Jan.	Order in C.	Great Britain. Renunciation. British Sovereignty. Orange Free State	174	814
	23 Feb.	Convention.	Great Britain and Orange Free State. Independence. Basutoland	174	814
	28 Sept.	Agreement.	Great Britain and Epé. Lagos. Palma.	92	406
1856.	12 July	Charter	Colony of Natal. Separate Colony.	93	435
1857.	7 Mar.	Convention.	Great Britain and France. Portendic and Albreda	106	544
1858.	5 Feb.	Order in C.	Great Britain. Boundaries. Natal.	93	436
	5 June	Proclamation.	Great Britain. Boundaries. Natal.	93	436

CHRONOLOGICAL LIST.

				No.	Page
1858.	17 June	Oath......	Krepee. British Allegiance	91	388
1859.	10 Feb.	Treaty.....	Sardinia and Ethiopia. Trade	4	4
	24 Aug.	Convention.	Spain and Morocco. Melilla, &c. ...	186	894
1860.	26 April	Treaty.....	Spain and Morocco. Ceuta, &c......	187	897
1861.	2 April	Award.....	Great Britain. Independence. Muscat and Zanzibar..................	205	961
	2 April	Treaty.....	Great Britain and Quiah. Cession. British Quiah	100	505
	11 June	Treaty.....	Great Britain and Ma Bwetie and Ra Wollah. Kambia. Boundaries ..	100	506
	22 June	Decision ...	Great Britain. Lagos. British Dependency......................	92	406
	6 Aug.	Treaty.....	Great Britain and Lagos. Cession to Great Britain. Port and Island of Lagos.........................	92	409
	6 Aug.	Proclamation.	Great Britain. Occupation. Lagos.	92	410
	30 Oct.	Treaty.....	Spain and Morocco. Limits. Melilla.	188	901
	9 Nov.	Treaty.....	Great Britain and Sherbro. Cession. Sherbro and Turtle Islands	100	507
	9 Nov.	Treaty.....	Great Britain and Bendoo and Chah. Cessions. Sherbro	100	508
	9 Nov.	Treaty.....	Great Britain and Bagroo. Cessions. Bagroo, &c......................	100	509
1862.	1 Feb.	Treaty.....	Great Britain and Quiah. British Quiah	100	510
	10 Mar.	Declaration.	Great Britain and France. Independence. Muscat and Zanzibar	107	547
	11 Mar.	Treaty.....	France and Danakils. Cession. Obock to France	68	269
1863.	7 Feb.	Declaration.	Epé. Lagos. Palma and Leckie....	92	411
	— Mar.	Conditions.	Peace. Lagos and Epé	92	411
	27 June	Agreement.	Great Britain and Addo. Protection.	92	412
	29 June	Agreement.	Great Britain and Pocrah. Protection	92	413
	4 July	Agreement.	Great Britain and Okeadan. Protection........................	92	413
	7 July	Agreement.	Great Britain and Badagry. Cession. Badagry to Great Britain........	92	414
	17 July	Deed......	Okeadan (Lagos). Sale. Land to Great Britain	92	415
	10 Oct.	Convention.	Great Britain and Tunis. Real Property.........................	190	906
	9 Dec.	Letters Pat.	Great Britain. Annexation. Nomansland, Natal	93	436
1864.	— Oct.	Oath......	Accoomfee. British Allegiance......	91	388
1865.	17 Mar.	Act	Incorporation of British Kaffraria with Cape Colony...............	89	343
	31 May	Convention.	Morocco and Foreign Powers. Cape Spartel Lighthouse	172	808
	— May	Firman	Turkey. Ports of Massowah and Suakin.........................	67	260
	27 June	Treaty.....	Great Britain and Madagascar. Privileges...........................	169	796

CHRONOLOGICAL LIST.

				No.	Page
1865.	7 Sept.	Proclamation.	Great Britain. Annexation. No-mansland to Natal	93	437
1866.	19 Feb.	Commission.	Union. Gambia, Gold Coast, Lagos, and Sierra Leone	100	511
	3 April	Treaty	Basutoland and Orange Free State	87	329
	27 May	Firman	Turkey. Massowah and Suakin, &c.	67	260
1867.	5 Mar.	Convention.	Great Britain and Netherlands. Exchange of Territories. Gold Coast.	137	674
1868.	12 Mar.	Proclamation.	Basutos. British Subjects. British Territory	87	330
	19 May	Treaty	Cession. Kotonou to France	65	249
	8 Aug.	Treaty	France and Madagascar	165	788
	30 Nov.	Treaty	Awoonah and Addah. Volta River	91	389
1869.	13 Jan.	Protocol	Great Britain and Portugal. Arbitration. Dispute. Bulama	139	684
	12 Feb.	Convention.	Great Britain and Orange Free State. Boundaries. Basutoland	174	814
	29 July	Treaty	Portugal and Transvaal. Boundary. Delagoa Bay. Lorenço Marquez	175	822
1870.	21 April	Award	United States. Sovereignty. Portugal. Bulama	140	688
	1 Oct.	Notification.	Portugal. Occupation. Bulama	141	691
1871.	25 Feb.	Convention.	Great Britain and Netherlands. Transfer to Great Britain. Dutch Possessions. Gold Coast	137	676
	11 Aug.	Act	Annexation. Basutoland to Cape Colony	87	330
	17 Oct.	Award	South African Republic and Bechuanaland. Boundary	179	840
	23 Oct.	Firman	Turkey. Boundaries. Tunis	190	906
	2 Nov.	Protocol	Great Britain and Netherlands. Free Labourers. Coast of Guinea	137	678
1872.	29 Jan.	Treaty	Great Britain and Quinh. Retrocession. Portions. British Quiah.	100	512
	25 Sept.	Protocol	Great Britain and Portugal. Arbitration. Dispute. Delagoa Bay	142	693
1873.	8 June	Firman	Turkey. Massowah, Suakin, &c.	67	260
	Sept.	British Case.	Dispute. Portugal. Delagoa Bay	143	694
1874.	13/14 Feb.	Treaty	Great Britain and Ashantee. Elmina, &c.	91	390
	— Mar.	Engagements.	Appolonia, &c. British Laws	91	390
	22 June	Treaty	Awoonah, Jellah Coffee, Quittah, &c.	91	391
	24 July	Letters Pat.	Great Britain. Gold Coast and Lagos. Gold Coast Colony	92	417
	17 Dec.	Letters Pat.	West African Settlements	100	513
	—	Act	Annexation of Ichaboe and Penguin Islands to Cape Colony	89	345
1875.	June	Agreement.	Great Britain and Portugal. Noncession. Delagoa Bay	144	697
	19 July	Convention.	Great Britain and Tunis. Consular Jurisdiction	190	907

CHRONOLOGICAL LIST.

				No.	Page
1875.	24 July	Award.....	France. Sovereignty. Portugal. Delagoa Bay........................	145	701
	6 Aug.	Order in C..	British Jurisdiction. Territories adjacent to Gold Coast.............	91	392
	11 Dec.	Treaty.....	Portugal and South African Republic. Lorenço Marquez Railway........	176	823
	21 Dec.	Treaty.....	Great Britain and Sherbro and Mendi. Customs Dues...................	100	514
	30 Dec.	Treaty.....	Great Britain and Sherbro and Mendi. Cessions........................	100	514
1876.	1 Mar.	Treaty.....	Liberia and Cape Palmas, &c. Submission of Chiefs................	164	778
	10 June	Treaty.....	Great Britain and Ro Wollah, &c. British Sovereignty. Scarcies Rivers.	100	514
	12 June	Letters Pat.	Annexation of Fingoland, Idutywa Reserve, &c., to Cape Colony.......	89	343
	13 July	Agreement..	Great Britain and Orange Free State. Boundaries.....................	174	818
	1848–1876	Notes......	On the Orange Free State..........	174	814
1877.	12 April	Proclamation.	Great Britain. Transvaal. British Territory.....................	179	840
	2 May	Treaty.....	Great Britain and Samoo Bulloms. Cession to Great Britain	100	516
	7 Sept.	Agreement..	Great Britain and Egypt. Somali Coast............................	67	260
1878.	12 Mar.	Proclamation.	British occupation of Walfisch Bay..	89	358
	19 April	Treaty.....	Cession. Kotonou to France.......	65	240
	4 Sept.	Proclamation.	Annexation. St. John's River to Cape Colony........................	89	347
	14 Dec.	Letters Pat.	British occupation of Walfisch Bay..	89	360
1879.	2 Aug.	Firman....	Turkey. Egypt. Non-cession of Rights, &c.......................	67	262
	24 Sept.	Agreement..	Great Britain and Katanu. British Protectorate....................	91	392
	29 Sept.	Proclamation.	Great Britain. Transvaal..........	179	841
	28 Oct.	Treaty.....	Zanzibar and Portugal. Consular Jurisdiction.......................	206	963
	8 Nov.	Letters Pat.	Great Britain. Transvaal..........	179	841
	1 Dec.	Agreement .	Afflowhoo. Cession. Seaboard to Great Britain....................	91	393
	2 Dec.	Agreement .	Agbosomé. British Jurisdiction. Seaboard........................	91	393
	6 Dec.	Agreement .	Afflowhoo. British Jurisdiction	91	394
1880.	25 Dec.	Notice.....	France. Limits. Obock..........	69	272
	1877–1880	Act, Order in C., and Proclamation.	Annexation. Griqualand West to Cape Colony....................	89	361
1881.	12 May	Treaty.....	France and Tunis. Temporary Occupation........................	190	907
	16 May	Protest.....	Turkey. Against Treaty, France and Tunis. 1881................	190	910
	16/20 May	Notes......	France and Tunis. Bizerta. Treaty Rights. Foreign Powers.........	108	548

CHRONOLOGICAL LIST.

Date		Type	Description	No.	Page
1881.	3 Aug.	Convention.	Great Britain and Transvaal. British Suzerainty. Swaziland, &c.	179	841
	30 Nov.	Decree	Egypt. Eastern Soudan	67	262
1882.	30 Mar.	Agreement..	Great Britain and Gallinas. Cessions to Great Britain.	100	518
	28 June	Convention.	Great Britain and France. Limits. North of Sierra Leone.	109	554
	Aug.	Conditions..	Surrender. Cetawayo. Zululand	102	532
	7 Oct.	Protocol....	Great Britain and Portugal. British Ratification. Treaty. Portugal and South African Republic. 1875	146	704
	18 Nov.	Agreement..	Great Britain and Bulloms and Shebar Cessions in 1825	100	520
1883.	15 Mar.	Treaty	Italy and Assab (Danakils). Cession.	5	5
	21 May	Treaty	Italy and Shoa. Boundaries, &c.	6	6
	5 June	Agreement..	Great Britain and Massah and Krim. Cessions	100	522
	21 Oct.	Agreement..	Great Britain and Krim. Cessions..	100	523
		Act	Cape. Disannexation. Basutoland.	87	331
	1863-1883	Notes	On Tunis	190	906
1884	2 Feb.	Order in C..	British direct authority over Basutoland	87	332
	26 Feb.	Treaty	Great Britain and Portugal. Boundary. West Coast of Africa. Rivers Congo and Zambesi	147	713
	27 Feb.	Convention.	Great Britain and South Africa Republic. Boundaries. Swaziland, &c...	179	847
	15 Mar.	Proclamation.	Great Britain. Protectorate over Katanu	92	417
	Feb.-Mar.	Agreement..	Treaties. Transvaal. Netherlands and Portugal	179	857
	9 April	Treaty	France and Gobad. Friendship, &c..	70	273
	22 April	Declaration.	United States and Congo. Recognition	64	244
	3 May	Treaty	Great Britain and Batlapings. Bechuanaland	88	334
	17 May	Treaty	Portugal and South Africa Republic. Lorenço Marquez Railway	176	827
	22 May	Treaty.	Great Britain and Barolongs. Bechuanaland	88	334
	Apr.-May	Notes	France and Congo. Rights of Preemption	46	207
	3 June	Treaty	Great Britain and Ethiopia and Egypt. Bogos, &c.	3	2
	5 July	Agreement..	Germany and Togoland. Protection.	84	320
	12 July	Proclamation.	Germany and Cameroons. Protection.	84	320
	16 July	Treaty	Great Britain and Jakri. Protection.	92	417
	19 July	Notification.	Great Britain. Ambas Bay. British Sovereignty	84	321
	22 July	Act	Annexation. St. John River Territories to Cape Colony	89	351
	22 July	Act	Annexation. Walfisch Bay to Cape Colony	89	360

CHRONOLOGICAL LIST.

Year	Date	Type	Description	No.	Page
1884.	7 Aug.	Proclamation.	Annexation. Walfisch Bay to Cape Colony	89	360
	5 Sept.	Notification.	German Protectorates. S.W. Coast of Africa	80	360
	21 Sept.	Treaty	France and Tajourah. Cession. Gubbed Kharab, &c.	71	274
	25 Sept.	Treaty	Little Popo. British Protection	91	394
	2 Oct.	Letters Pat.	Annexation. Tembuland. Emigrant Tembuland, Galekaland, and Bomvanaland to Cape Colony	89	362
	15 Oct.	Notification.	Germany. Protectorate, West and South West Coast of Africa	84	320
	18 Oct.	Act	Tajourah. Cession to France. Gubbed Kharab, &c.	72	276
	8 Nov.	Declaration.	Germany and Congo. Recognition	52	219
	14 Dec.	Act	Tajourah. Cession to France. Adaeli to Ambado	73	277
	16 Dec.	Declaration.	Great Britain and Congo. Recognition	53	221
	16 Dec.	Convention.	Great Britain and Congo. Consular Jurisdiction, &c.	54	223
	18 Dec.	Notification.	Great Britain. Occupation. St. Lucia Bay	93	437
	19 Dec.	Convention.	Italy and Congo. Recognition	55	227
	24 Dec.	Declaration.	Austria-Hungary and Congo. Recognition	40	194
	24 Dec.	Treaty	Great Britain and Ogbo. Protection.	92	418
	27 Dec.	Convention.	Netherlands and Congo. Recognition.	57	230
	Nov.–Dec.	Treaties	German East African Society and Native Chiefs	81	303
	1882–1884	Treaties	Belgian Expedition. Native Chiefs. Upper Congo	—	200
1885.	5 Jan.	Notification.	British Protectorate. Coast of Pondoland	89	351
	7 Jan.	Convention.	Spain and Congo. Recognition	62	240
	9 Jan.	Notification.	Spain. Protectorate. North-West Coast of Africa	182	886
	27 Jan.	Order in C.	Great Britain. Bechuanaland. Kalahari	88	334
	3 Feb.	Proclamation.	Italy. Occupation. Massowah	7	8
	5 Feb.	Convention.	France and Congo. Recognition	47	209
	5 Feb.	Convention.	France and Congo. Private Stations and Properties	48	212
	5 Feb.	Convention.	Russia and Congo. Recognition	61	239
	10 Feb.	Convention.	Sweden and Norway and Congo. Recognition	63	242
	11 Feb.	Notification.	France. Protectorate. Ras Ali to Gubbet Kharab	71	274
	14 Feb.	Convention.	Portugal and Congo. Recognition	58	232
	17 Feb.	Charter	German Colonisation Society	81	303
	23 Feb.	Declaration.	Belgium and Congo. Recognition	41	196
	23 Feb.	Convention.	Denmark and Congo. Recognition	45	205

CHRONOLOGICAL LIST.

				No.	Page
1885.	26 Feb.	"Berlin Act."	Trade and Civilization in Africa. Rivers Congo, Niger, &c. Slave Trade. Occupation, &c., African Coasts	17	20
	18 May	Agreement..	Great Britain and Gallinas. Boundaries	100	524
	28 May	Treaty	Zanzibar and Italy. Consular Jurisdiction	201	{945 948}
	30 May	Treaty	Zanzibar and Belgium. Consular Jurisdiction, &c.	195	926
	1 June	Treaty	National Africa Company and Sokoto. (1) Jurisdiction over Foreigners, &c.	App. 1	972
	5 June	Notification.	Great Britain. Protectorate. Niger Districts	95	445
	13 June	Treaty	National Africa Company and Gandu. (1) Transfer of Rights, &c.	App. 2	974
	17 June	Letters Pat..	British West Africa Settlements	100	525
	Apr.-June	Arrangement.	Great Britain and Germany. Spheres of Action. Gulf of Guinea, &c.	119	596
	May-June	Treaties	East African Society and Native Chiefs	81	{303 306}
	25 July	Notification.	Great Britain. St. Lucia Bay	93	437
	1 Aug.	Circular	Congo. Neutrality. Limits	42	198
	5 Aug.	Award	Orange Free State. Boundary Dispute. Great Britain and South African Republic	179	858
	5 Aug.	Treaty	Dahomey and Portugal. Portuguese Protectorate	65	253
	29 Sept.	Commission.	Government. Bechuanaland	88	334
	30 Sept.	Proclamation.	Cape. Bechuanaland, Kalahari, &c.	88	335
	24 Oct.	Treaty	Great Britain and Mahin. Cession. Mahin Beach	92	418
	24 Oct.	Treaty	Great Britain and Mahin. Protection. Atijere	92	421
	11 Nov.	Convention.	Great Britain and Liberia. Boundaries	164	781
	12 Nov.	Treaty	Krikor. Cession to Great Britain	91	395
	22 Nov.	Protocol	France and Congo. Manyanga	49	213
	17 Dec.	Treaty	France and Madagascar. Foreign Relations	106	793
	20 Dec.	Treaty	Zanzibar and Germany. Consular Jurisdiction	197	930
	24 Dec.	Protocol	France and Germany. Biafra Bay, Togoland, Senegambia, &c.	78	293
	27 Dec.	Declaration.	France. Treaties. Madagascar and Foreign Powers	107	794
	1750-1885	Treaties	France and Native Chiefs. Cession. Islands, &c. Madagascar	168	795
	1843-1885	Notes	On Natal	93	434
1886.	13 Jan.	Letters Pat..	Lagos. Separate Colony	92	422
	13 Jan.	Order in C..	Gold Coast Settlements. Separate Colony	91	396

CHRONOLOGICAL LIST.

				No.	Page
1886.	21 Jan.	Notification.	Portuguese Protectorate. Dahomey.	65	253
	5 Feb.	Proclamation.	Great Britain. Sovereignty. Odi to Benin River	92	422
	19 April	Ratifications.	Berlin Act	17	45
	21 April	Convention.	France and Johanna. Protection.	76	291
	23 April	Treaty	Great Britain and Socotra. Protectorate	177	828
	30 April	Treaty	Great Britain and Zanzibar. Ex-territoriality. Consular Jurisdiction, &c.	152	751
	5 May	Declaration.	Badagry. Town of Badagry	92	424
	12 May	Convention.	France and Portugal. Limits. Guinea, &c.	80	298
	9 June	Procès-Verbal.	Great Britain, France, and Germany. Zanzibar Limits	120	605
	26 June	Notification.	France. Protectorate. Comoro Islands.	77	292
	3 July	Convention.	Zanzibar and United States. Consular Jurisdiction, &c.	208	966
	10 July	Royal Charter.	Great Britain. National African Company	96	416
	15 July	Protocol	Great Britain and Germany. British Claims. German Protectorates. South-West Africa	121	608
	23 July	Declaration.	Dakami. Frah. Denham Waters. Whemi	92	424
	27 July	Treaty	Aquamoo (Volta). Incorporation with Gold Coast	91	396
	12 Aug.	Treaty	Aggravie (Volta). Cession to Great Britain	91	396
	23 Aug.	Letters Pat.	Annexation. Xesibe Country, Mt. Ayliff, to Cape Colony	89	351
	July—Aug.	Arrangement.	Great Britain and Germany. Spheres of action. Gulf of Guinea, &c.	122	612
	4 Sept.	Treaty	Battor, &c. (Volta). Cession to Great Britain	91	397
	22 Oct.	Agreement.	Great Britain and New Republic. Zululand Boundary	179	860
	25 Oct.	Proclamation.	Annexation. Xesibe Country to Griqualand East	89	352
	29 Oct.	Accession	Germany. Anglo-French Agreement, 1862. Zanzibar	123	618
	8 Nov.	Accession	Zanzibar. "Berlin Act"	193	925
	Oct.—Nov.	Treaties	Great Britain and Crepee. Incorporation with Gold Coast	91	397
	Oct.—Nov.	Agreement.	Great Britain and Germany. Zanzibar Limits, &c.	123	615
	4 Dec.	Adhesion	Zanzibar. Anglo-German Agreement. Zanzibar Limits	124	622
	3—4 Dec.	Agreement.	Great Britain and Zanzibar. Zanzibar Limits	153	754
	9 Dec.	Agreement.	Pondoland, &c. Cessions to Cape Colony	89	354
	30 Dec.	Declaration.	Germany and Portugal. South-West and South-East Africa	85	323

CHRONOLOGICAL LIST.

				No.	Page
1886.	1884–1886	Treaties....	Great Britain and Native Tribes. Somali Coast. Protectorate......	178	834
1887.	1 Feb.	Procès-Verbal.	France and Germany. Slave Coast ..	79	297
	18 Feb.	Treaty.....	Great Britain and Schwhi. British Protection......................	91	397
	Mar.	Agreement..	Great Britain and Germany. Trading Stations. Spheres of Influence....	125	623
	6 April	Decree.....	Spain. North-West Coast of Africa..	182	887
	22—29 April	Notes......	France and Congo. Right of Preemption	50	215
	29 April	Protocol....	France and Congo. Oubangi........	51	217
	9 May	Declaration.	Aquamoo (Volta). British Protection.	91	397
	14 May	Proclamation.	Great Britain. Zululand a British Possession......................	102	533
	24 May	Concession..	Zanzibar to British East Africa Company	24	110
	June	Treaties....	British East Africa and Native Chiefs	—	164
	6 July	Agreement..	Great Britain and Amatonga. Boundaries, &c........................	101	529
	7 July	Convention.	Italy and Danakils. Aussa.........	8	8
	20 July	Notification.	British Protectorate. Somali Coast. Ras Jiburti to Bunder Ziadeh......	178	834
	25 July	Treaty.....	Betente. Cession to Great Britain ..	90	381
	29 July	Letters Pat..	Annexation. Rodo Valley. Cape Colony........................	89	357
	30 July	Declaration.	Gama and Bontuku. Allegiance. British Crown..................	91	398
	30 July	Treaty.....	South African Republic and Matabeleland	170	861
	July	Agreement..	Great Britain and Germany. Nonannexation in rear of Spheres of Influence.......................	126	625
	10 Aug.	Convention.	Italy and Danakils. Aussa Road....	9	9
	11 Aug.	Treaty.....	Zanzibar and Austria-Hungary. Consular Jurisdiction, &c.	194	926
	13 Aug.	Protest	Great Britain. Treaties. Portugal and France, and Portugal and Germany	85	325
	14 Sept.	Treaty.....	South African Republic and New Republic Union..................	170	862
	15 Sept.	Treaty.....	Jolah. Cession to Great Britain....	90	382
	17 Sept.	Treaty.....	Vintang (Fogni). Cession to Great Britain	90	382
	19 Sept.	Treaty.....	Central Kiang. Cession to Great Britain	90	383
	11 Oct.	Treaty.....	Jarru. Cession to Great Britain	90	383
	15 Oct.	Convention.	France and Johanna. Protectorate..	77	292
	18 Oct.	Notification.	Great Britain. Protectorate. Niger Districts.......................	97	449
	20 Oct.	Convention.	Italy and Shoa. Alliance..........	10	9
	22 Dec.	Notification.	Portuguese Protectorate. Dahomey. Withdrawn....................	65	253

CHRONOLOGICAL LIST.

				No.	Page
1887.	20 Dec.	Order in C..	British Jurisdiction. Territories adjacent to Lagos (S.P. 78, p. 836) ...		
	20 Dec.	Order in C..	British Jurisdiction. Territories adjacent to Gold Coast	91	398
	Dec.	Report.....	British and German Commissioners. Interior of Gold Coast, Togoland, &c...........................	127	628
	Dec.	Notes......	Spain and Italy. Spanish Naval Station. Assab Bay................	183	888
1888.	2—9 Feb.	Agreement..	Great Britain and France. Gulf of Tajourah, &c. Somali Coast.......	App. 3	976
	28 April	Concession..	Zanzibar to German East Africa Company. Mrima and South of River Umba	198	933
	5 May	Treaty	Quahoe. British Protection........	91	398
	15 May	Declaration .	Igbessa. Boundaries, &c...........	92	425
	15 May	Proclamation.	Great Britain. Protectorate. Igbessa.	92	426
	22 May	Declaration.	Ife. Boundaries, &c.	92	427
	28 May	Declaration.	Itebu. Boundaries, &c. ......;....	92	427
	29 May	Declaration.	Ketu. Boundaries, &c.	92	427
	29 May	Treaty	Great Britain and Ketu. Protectorate, Ketu	92	427
	31 May	Declaration.	Ibu. Boundaries, &c.	92	428
	4 June	Agreement..	Zanzibar and German East Africa Company. Custom Houses, &c. ...	199	941
	20 June	Convention.	Great Britain and South African Republic. Union. New Republic and South African Republic..........	179	862
	June	Declaration .	Kotoku. Insuaim Ferry. Berim River	91	398
	3 July	Declaration .	Agotine (Crepee). Allegiance. British Crown.........................	91	399
	3 July	Declaration .	Anum and Buem. Fealty. British Crown.........................	91	399
	3 July	Declaration .	Crepee. British Protection........	91	399
	3 July	Declaration .	Aduklu, &c. (Crepee). Allegiance to Head King	91	400
	21 July	Convention.	Great Britain and Ilaro. Boundaries, &c.	92	429
	21 July	Declaration.	Great Britain. Protectorate. Ilaro..	92	429
	23 July	Treaty	Great Britain and Oyo and Yorubaland. Boundaries................	92	430
	1 Aug.	Decree......	Congo Administrative Districts......	43	201
	2 Aug.	Notification .	Italy. Protectorate over Zula	11	10
	17 Aug.	Declaration..	Great Britain. Abolition. Consular Jurisdiction. Massowah	134	664
	3 Sept.	Charter.....	British East Africa Company........	25	118
	9 Oct.	Concession ..	Zanzibar to British East Africa Company...........................	26	125
	9 Oct.	Memorandum.	Artijere Wharf. Boundaries	92	431
	14 Oct.	Treaty	Batelling (Central Kiang). Cession to Great Britain	90	383

CHRONOLOGICAL LIST.

				No.	Page
1888.	28 Nov.	Letters Pat.	Colony of Sierra Leone	101	526
	28 Nov.	Letters Pat.	Gambia. Separate Colony	90	383
	9 Dec.	Treaty	Italy and Danakils. Italian Sovereignty	12	11
	9 Dec.	Notification.	Great Britain. Boundaries. Zululand	102	534
	1843–1888	Notes	On Zululand	102	532
1889.	20 Feb.	Treaty	Great Britain and Ondo. Boundaries	92	432
	2 Mar.	Notification.	Italy. Protectorate over Oppia	161	772
	2 May	Treaty	Ethiopia and Italy. Boundaries. Foreign Relations	13	12
	20 May	Notification.	Italy. Protectorate over Oppia	162	774
	3 Aug.	Agreement..	British East Africa Company and Italy. Kismayu. Benadir Ports	27	137
	10 Aug.	Arrangement.	Great Britain and France. Senegambia to Gold Coast. Slave Coast. Gambia. Sierra Leone. Porto Novo, &c.	110	558
	17 Aug.	Award	Belgium. Differences. British East Africa Company and German Witu Company. Island of Lamu, &c.	128	630
	31 Aug.	Agreement..	Great Britain and Zanzibar. Lease. British East Africa Company	154	760
	1 Oct.	Convention.	Ethiopia and Italy. Boundaries, &c...	14	15
	12 Oct.	Notification.	Italy. Ethiopian Foreign Affairs	15	17
	22 Oct.	Notification.	Germany. Protectorate over Witu, &c.	81	315
	29 Oct.	Charter	British South Africa Company	37	174
	18 Nov.	Deed	Transfer. British East Africa Company to Italy. Benadir Ports	28	142
	19 Nov.	Notification.	Italy. Protectorate over Part of East African Coast	163	776
	2–19 Nov.	Notes	Great Britain and France. Approval Boundary Arrangement 10 August, 1889	111	568
	6 Dec.	Notification.	Italy. Protectorate over Aussa (Danakils)	16	18
	13 Dec.	Order in C.	Great Britain. Jurisdiction. Somali Coast	178	835
	16 Dec.	Law	Turkey. Black Slaves	192	919
	21 Dec.	Agreement..	Zanzibar and British East Africa Company. Customs Dues. Wanga to Kipini	29	146
	1782–1889	Notes	On the Gambia	90	365
	1840–1889	Notes	On the Somali Coast	178	832
1890.	13 Jan.	Agreement..	Zanzibar and German East Africa Company. Custom Houses	200	943
	20 Jan.	Treaty	Royal Niger Company and Boussa (Borgu) British Protection	App. 4	981
	4 Mar.	Concession..	Zanzibar. British East Africa Company	30	148

CHRONOLOGICAL LIST.

				No.	Page
1890.	5 Mar.	Letter	Zanzibar. British East Africa Company and Italy. Benadir Ports....	—	140
	7 April	Treaty	Royal Niger Company and Gandu (2) Protection, &c.	App. 5	983
	8 April	Notification.	Italy. Zanzibar and Italy. Benadir Ports, &c.	202	940
	15 April	Treaty	Royal Niger Company and Sokoto (2). Jurisdiction, &c.	App. 6	984
	10 June	Decree	Congo. Administrative Districts. Eastern Kwango	44	204
	14 June	Agreement..	Great Britain and Zanzibar. British Protectorate	155	703
	30 June	Order in C...	Great Britain. Territories North of British Bechuanaland.	88	336
	1 July	Agreement..	Great Britain and Germany. Spheres of Influence. East, West, and South-West Africa	129	642
	2 July	"Brussels Act"	Slave Trade, &c..	18	48
	,,		Ditto. Import Duties	18	88
	2 July	Accession...	Persia. "Brussels Act"	Note	48
	5 Aug.	Declaration.	Great Britain and France. British Protectorate. Zanzibar and Pemba.	112	570
	5 Aug.	Declaration.	Great Britain and France. French Protectorate. Madagascar. Sphere of Influence. Saye to Barruwa....	113	571
	20 Aug.	Convention.	Great Britain and Portugal. Spheres of Influence. Zambesi, &c.	148	715
	July—Aug.	Convention.	Great Britain and South African Republic. Swaziland	179	868
	16 Sept.	Accession ...	Ethiopia. "Brussels Act"	Note	48
	3 Oct.	Arrangement.	France and Dahomey. Protectorate. Porto Novo.	65	253
	4 Oct.	Concession.	Gazaland to British South Africa Company.	37	184
	4 Nov.	Notification.	Great Britain. Protectorate. Zanzibar.	156	766
	14 Nov.	Agreement..	Great Britain and Portugal. Spheres of Influence. Zambesi, &c.	149	728
	17 Nov.	Agreement..	France and Germany. Madagascar, Zanzibar, and Mafia.	App. 7	985
	19 Nov.	Notification .	Great Britain. Protectorate. Witu, &c.	86	327
	22 Dec.	Agreement..	Great Britain, Germany, and Italy. Tariff. Eastern Zone. Congo Basin.	19	90
	1819-1890	Treaties....	France and African Chiefs.	74	278
	1852-1890	Notes	On the South African Republic (Transvaal)....	179	839
	1881-1890	Notes	On Swaziland.	189	903
	1884-1890	Notes	On German Protectorates. East Africa.	82	305
	1884-1890	Notes	On German Protectorates. West Coast of Africa.	84	320

CHRONOLOGICAL LIST.

				No.	Page
1891.	25 Jan.	Terms of Peace.	Great Britain and Witu............	—	156
	2 Feb.	Declaration.	Great Britain and Zanzibar. Jurisdiction........................	157	767
	6 Feb.	Agreement..	Italy and Ethiopia. Boundary......	Note	16
	Feb.	Conditions..	Extension of operations of British South Africa Company to the north of the Zambesi..................	App. 8	987
	5 Mar.	Agreement..	Zanzibar and British East Africa Company. Benadir Ports............	31	150
	5 Mar.	Agreement..	British Government and British East Africa Company. Witu..........	32	153
	18 Mar.	Agreement..	British East Africa Company and Witu...........................	33	157
	20 Mar.	Notice	British East Africa Company. Administration. Witu and Kipini to Kismayu......................	34	160
	24 Mar.	Protocol....	Great Britain and Italy. Spheres of Influence. Juba to Blue Nile	135	665
	2 April	Agreement .	British Government and British South Africa Company. Operations North of the Zambesi............	37	185
	13 April	Proclamation.	Bechuanaland. Territory North of the Limpopo....................	88	336
	15 April	Protocol....	Great Britain and Italy. Spheres of Influence. Ras Kasar to Blue Nile	136	667
	8 May	Proclamation.	Annexation. Bastards or Griqua Country to Bechuanaland	88	336
	9 May	Order in C.	Great Britain. British Jurisdiction north of British Bechuanaland	38	183
	14 May	Notification.	Great Britain. Protectorate. Nyasaland	173	811
	25 May	Treaty.....	Congo and Portugal. Boundaries. Lunda Region	59	234
	25 May	Convention.	Congo and Portugal. Boundaries. Lower Congo	60	236
	11 June	Treaty.....	Great Britain and Portugal. Spheres of Influence. East and Central Africa.	150	731
	26 June	Agreement .	Great Britain and France. Spheres of Influence. Niger Districts	114	573
	27 June	Proclamation.	Cape. Jurisdiction. Protected Territories. Bechuanaland, Tati District, &c.	App. 9	990
	2 July	Protocol ...	Ratifications. "Brussels Act"....	20	92
	5 Aug.	Note	Igbessa added to Lagos............	92	425
	8 Aug.	Note	Addo added to Lagos	92	412
	13 Aug.	Note	Ilaro added to Lagos	92	429
	26 Oct.	Notification.	France. Protectorate. Ivory Coast..	164	783
	15 Dec.	Treaty.....	Congo and Liberia. Most favoured Nation Treatment	56	229
	1819-1891	Treaties ...	France and African Chiefs	74	278
	1843-1891	Notes......	Great Britain. Basutoland........	87	329
	1831-1891	Notes......	Great Britain. Gold Coast........	91	388

CHRONOLOGICAL LIST.

				No.	Page
1891.	1851-1891	Notes	Great Britain. Lagos	92	405
	1885–1891	Treaties	Great Britain and Native Chiefs. Sierra Leone	100	526
	1887-1891	Treaties	British East Africa Company and Native Chiefs	35	164
	1887-1891	Notes	Boundaries. British Spheres of Influence. East Coast of Africa	36	174
	1889-1891	Treaties	British South Africa Company and Native Chiefs	39	187
	1890-1891	Notes	On the British South Africa Company.	38	183
1892.	2 Jan.	Protocol	Ratifications. "Brussels Act"	21	97
	2 Feb.	Protocol	Ratification. United States. "Brussels Act"	22	102
	8 Feb.	Notice	Great Britain. Free Port. Zanzibar.	App. 10	992
	27 Mar.	Firman	Turkey. Boundaries. Egypt	67	205
	30 Mar.	Protocol	Ratification. Portugal. "Brussels Act"	23	104
	30 Mar.	Treaty	British East Africa Company and Uganda (not ratified)	—	172
	18 June	Notification	Great Britain. Alcoholic Liquors. Witu. Niger Protectorate. Nyasaland	86	328
	22 June	Notification	Great Britain. Zanzibar. Free Trade Zone. "Berlin Act"	App. 11	993
	12 Aug.	Treaty	Zanzibar and Italy	203	950
	25 Aug.	Accession	Liberia. "Brussels Act"	18	48
	3 Dec.	Proclamation	French Protectorate. Dahomey	65	248
	6 Dec.	Genl. Order	France. Limits. Whydah	65	248
	8 Dec.	Arrangement	France and Liberia. Limits	164	783
	16 Dec.	Declaration	Great Britain and Zanzibar. Consular Jurisdiction	158	768
	Oct.—Dec.	Protocol	Great Britain and Germany. Limits. Wanga. Lake Jipé. Kilimanjaro	130	652
	1788-1892	Notes	On Sierra Leone	100	484
	1884-1892	Treaties	Great Britain. National Africa Company, Royal Niger Company, and Native Chiefs	98	450
	1891-1892	Notes	Great Britain and France. Ivory Coast	115	575
	1884-1892	Notes	German Protectorates. South-West Africa	83	317
1893.	22 Feb.	Notification	Great Britain. "Nyasaland," "British Central Africa Protectorate"	173	811
	14 April	Agreement	Great Britain and Germany. Gulf of Guinea. Rio del Rey	131	654
	8 May	Report	British and French. Panchang, &c. (Gambia)	117	588
	13 May	Notification	Great Britain. "Oil Rivers Protectorate," "Niger Coast Protectorate."	99	479
	15 May	Agreement	Zanzibar and Italy. Benadir Ports	204	958
	29 May	Agreement	British Commissioner and Uganda	App. 12	995
	May—June	Agreement	Great Britain and Portugal. North of the Zambesi	151	743

CHRONOLOGICAL LIST.

				No.	Page
1893.	12 July	Arrangement.	Great Britain and France. Gold Coast.	118	589
	17 July	Order in C.	Great Britain. Natives of British Protectorates in Zanzibar	159	769
	25 July	Agreement..	Great Britain and Germany. Wanga. Jipé. Kilimanjaro	132	656
	31 July	Proclamation.	Great Britain. Administration. British Protectorate, North of the Tana.	160	770
	15 Aug.	Treaty	Great Britain and Ibadan. Zomba Towns........................	92	432
	23 Sept.	Treaty	Liberia and Greboes. Submission...	164	786
	8 Nov.	Convention.	Great Britain and South African Republic	189	903
	10 Nov.	Treaty	Liberia and Cavalla. Submission....	164	786
	15 Nov.	Agreement..	Great Britain and Germany. Gulf of Guinea. Interior...............	133	658
	1479-1893	Notes	On Spanish Possessions in Africa	180	882
	1816-1893	Notes	On Liberia and Maryland	164	778
	1847-1893	Notes	On Dahomey, Kotonou, Whydah, Porto Novo, &c.	65	248
	1866-1893	List	Orders in Council. Zanzibar........	—	750
	1882-1893	Notes	On the Niger Districts and Niger Coast Protectorates	94	440
1894.	5 Jan.	Treaty	France and Dahomey. Submission.	App. 13	998
	29 Jan.	Treaty	French Protectorate. King of Abomey.	App. 13	998
	4 Feb.	Treaty	French Protectorate. King of Allada.	App. 13	998
	4 Feb.	Protocol....	France and Germany. Cameroons and French Congo. Lake Chad..	App. 14	999
	24 Feb.	Convention .	Great Britain and Germany. Customs Union. Gold Coast Colony. East of the Volta. Togoland	133	661
	5 Mar.	Treaty	Spain and Morocco. Events near Melilla, and Indemnity, &c., Oct., Nov., 1893	Note	902
	15 Mar.	Agreement ..	France and Germany. Boundaries. Cameroons. French Congo. Lake Chad	App. 14	999
	20 Mar.	Proclamation.	British Sovereignty over Pondoland..	89	358
	24 Mar.	Declaration .	Congo and Portugal. Boundary. Lunda Region	App. 15	1004
	April	Proclamation.	Great Britain. Annexation. Foreign Combo to the Garnbia	90	384
	5 May	Protocol....	Great Britain and Italy. Eastern Africa. Somali, Harrar, &c.	136*	669
	12 May	Agreement..	Great Britain and Congo. East and Central Africa. Leases, &c.....	App. 16	1008
	12 June	Notification.	Annexation of Pondoland to Cape Colony.......................	89	358
	18 June	Notification .	Great Britain. Protectorate. Uganda.	App. 17	1016
	22 June	Decree......	France. Dahomey Coast. French Colony	App. 13	998
	22 June	Declaration .	Great Britain and Congo. Withdrawal of Art. 3 of Agreement of 12th May, 1894	App. 18	1017

4 H

CHRONOLOGICAL LIST.

				No.	Page
1894.	4 July	Treaty	Royal Niger Company and Gandu (3). Jurisdiction over Foreigners, &c.	App. 19	1018
	18 July	Order in C.	Great Britain. Matabeleland. Boundaries, &c.	App. 20	1020
	14 Aug.	Agreement..	France and Congo Free State. Boundaries	App. 21	1021
	27 Aug.	Treaty	Great Britain and Uganda. British Protectorate	App. 22	1023
	Sept.	Agreement..	Germany and Portugal. Spheres of Influence. East Africa. Kionga, &c.	App. 23	1024
	24 Nov.	Agreement..	British South Africa Company. British Central Africa north of the Zambesi	App. 24	1025
	10 Dec.	Convention..	Great Britain and South African Republic. Swaziland	App. 25	1029
	1652–1894	Notes	On the Cape of Good Hope	89	341
	1840–1894	Notes	On Boundaries of Egypt	67	259
1895.	7 Jan.	Declaration.	Great Britain and Portugal, Arbitration. Boundary Dispute. Manica Plateau	App. 26	1037
	9 Jan.	Treaty*	Congo and Belgium. Cession	App. 27	1041
	11 Jan.	Declaration*.	Belgium. Neutrality. Congo State.	App. 28	1043
	21 Jan.	Agreement ..	Great Britain and France. Boundary. North and East of Sierra Leone	App. 29	1048
	5 Feb.	Arrangement.*	Belgium and France. French Right of Pre-emption over Territories of Congo State	App. 30	1059
	5 Feb.	Declaration*..	Belgium and France. Possessions in Stanley Pool	App. 31	1061
	24 Feb.	Suppl. Convention.	Spain and Morocco. Indemnities, &c.	App. 32	1062
	13 Mar.	Agreement ..	Great Britain and Morocco. Purchase of Property of North-West Africa Company in Terfaya (Cape Juby)	App. 33	1064
	23 April	Proclamation.	Natal. British Sovereignty over Territories of certain Native Chiefs in Zululand	App. 34	1067
	11 June	Notification ..	British Protectorate. Amatongaland	App. 35	1068
	15 June	Notification ..	British Protectorate. British East Africa	App. 36	1069
	25 June	Agreement ...	Italy and Egypt. Boundary between the Baraka and the Red Sea.	App. 37	1072
	Sept.-Oct.	Exchange of Notes.	Great Britain and Portugal. Boundary. Tongaland and Portuguese Possessions	App. 38	1075
	4 Nov.	Notification ..	Great Britain. Non-recognition of certain Concessions made to the Amatongaland Exploration Company	App. 39	1078

* Treaty 9th January, 1895, not yet ratified (February 1, 1896).

ALPHABETICAL INDEX.

ALPHABETICAL INDEX TO VOLS. I AND II, AND APPENDIX.

Name of Country, Place, &c.	Subject.	Date of Treaty or other Document.	No. of Doc.	Art.	Page.
Abassuen	British and French Spheres	2–9 Feb., 1888	App. 3	1	972 976
Ablis (Aussa)	Cession of Part to Italy	15 Mar., 1853	5	5	5
Abomey-Kalavy	(Dahomey). Annexed to France	3 Dec., 1892	65	—	248
,,	,, ,,	29 Jan., 1894	App. 13	—	998
Abyssinia	Shoa and Great Britain. Treaty. Friendship, &c.	16 Nov., 1841	1	—	2
,,	Abyssinia and Great Britain. Treaty. Friendship, &c.	2 Nov., 1849	2	—	2
,,	Ethiopia and Great Britain and Egypt. Treaty. Bogos, &c.	3 June, 1884	3	—	2
,,	Ethiopia and Sardinia. Treaty. Commerce	10 Feb., 1859	4	—	4
,,	Assab (Danakils) and Italy. Treaty. Cession. Abliss (Aussa), &c.	15 Mar., 1883	5	—	5
,,	Shoa and Italy. Treaty. Boundaries, &c.	21 May, 1883	6	—	6
,,	Italy. Proclamation. Occupation of Massowah	3 Feb., 1885	7	—	8
,,	Aussa (Danakils) and Italy. Convention. Road. Assab to Mt. Musalli	7 July, 1887	8	—	8
,,	Aussa (Danakils) and Italy. Convention. Assab-Aussa-Shoa Road	10 Aug., 1887	9	—	9
,,	Shoa and Italy. Convention. Alliance	20 Oct., 1887	10	—	9
,,	Italy. Protectorate over Zula. Notification	2 Aug., 1888	11	—	10
,,	Aussa (Danakils) and Italy. Treaty. Italian Protectorate	9 Dec., 1888	12	—	11

ALPHABETICAL INDEX.

Name of Country, Place, &c.	Subject.	Date of Treaty or other Document.	No. of Doc.	Art.	Page.
Abyssinia	Ethiopia and Italy. Treaty. Boundaries. Foreign Relations	2 May, 1889	13	—	12
,,	Ethiopia and Italy. Additional Convention. Boundaries, &c.	1 Oct., 1889	14	—	15
,,	Conduct of Ethiopian Foreign Affairs by Italy. Notification	12 Oct., 1889	15	—	17
,,	Italy. Notification. Protectorate. Aussa (Danakils)	6 Dec., 1889	16	—	18
,,	Confines. British and German Spheres of Influence. See also Italy.	1 July, 1890	129	1, § 2	644
Accra	British Jurisdiction	22 June, 1874	91	—	391
Acquisitions	African Coasts. To be notified to Powers	26 Feb., 1885	17	6, 34, 35	23–43
,,	Return. Notifications	1885—1887	—	—	47
Adaeli	Cession by Tajourah to France	14 Dec., 1884	73	—	277
Addah	Treaty. Great Britain	30 Nov., 1868	91	—	389
,, ,,	British Jurisdiction. See also Ahwoonah.	22 June, 1874	91	—	391
Addo	(Lagos). British Protectorate	27 June, 1863	92	—	412
,, ,,	River. Free Navigation	10 Aug., 1889	110	4, § 1	562
,, ,,	And Pocrah. Custom House	10 Aug., 1889	110	4, § 5	563
Aden	Gulf of. British and Italian Spheres	5 May, 1894	136*	1	669
Afflowhoo	British Jurisdiction over Seaboard	1, 6 Dec., 1879	91	—	393 394
Africa	(Central). British and Portuguese Limits	11 June, 1891	150	4	733
,, ,,	,, ,, See also Congo.	May—June, 1893	151	5	744
,, ,,	(East Coast).	—	—	—	105
,, ,,	,, Spheres of Influence. See Congo. Great Britain and France. Great Britain and Germany. Great Britain and Italy. Great Britain and Portugal, and Great Britain and Zanzibar. See also British East Africa Company.				

ALPHABETICAL INDEX.

Name of Country, Place, &c.	Subject.	Date of Treaty or other Document.	No. of Doc.	Art.	Page.
Africa	(General). "Berlin Act." Trade and Civilization. Free Navigation of Rivers Congo, Niger, &c. Slave Trade by Sea and Land. Occupation of Territory on African Coasts, &c. ..	26 Feb., 1885	17	—	20
,, ...	,, "Brussels Act." African Slave Trade, &c.	2 July, 1890	18	—	48
,, ...	,, Declaration. Import Duties ..	—	—	—	83
,, ...	,, Tariff. Eastern Zone. Congo Basin	22 Dec., 1890	19	—	90
,, ...	,, Protocol. Ratifications. "Brussels Act." Various	2 July, 1891	20	—	92
,, ...	,, Protocol. Ratifications. Do. Austria - Hungary, Russia, Turkey, France, Netherlands	2 Jan., 1892	21	—	97
,, ...	,, Protocol. Ratifications. Do. United States ..	2 Feb., 1892	22	—	102
,, ...	,, Protocol. Ratifications. Do. Portugal	30 Mar., 1892	23	—	104
,, ...	(North).	—	—	—	105
,, ...	(North-West). Concession of Land to British Subjects	19 April, 1879	—	—	105
,, ...	(South). .. .. .. See also Matabeleland.	—	—	—	105
,, ...	(South-West Coast). ..	—	—	—	105
,, ...	(South-West). British Claims in German Protectorate. See also Great Britain and Germany.	15 July, 1886	121	—	608
,, ...	(West Coast).	—	—	—	105
,, ...	,, British and French Boundary. See Great Britain and France.				
,, ...	,, British and German Boundary. See Germany, and Great Britain and Germany.				
African Coasts	Access of British and French Subjects ..	3 Sept., 1783	108	12	539

ALPHABETICAL INDEX.

Name of Country, Place, &c.	Subject.	Date of Treaty or other Document.	No. of Doc.	Art.	Page.
African Coasts	Occupation, Protectorates, &c. To be notified to Powers	26 Feb., 1885	17	6, 34, 35	23, 43
,, ,,	Maritime Zone defined	26 Feb., 1885	17	1, § 2	24
,, ,,	Free Access. All Flags	26 Feb., 1885	17	2	25
Africa Company	Abolition. Transfer of Possessions, &c., to British Government	7 May, 1821	100	—	488
African Lakes Company	War Material. Steamers	Feb., 1891	App. 8	8	989
,, ,,	,, ,,	24 Nov., 1894	App. 24	3	1026
,, ,,	Note	24 Aug., 1891	38	—	186
,, ,,	Land and Minerals claimed by British South Africa Company	24 Nov., 1894	App. 24	7	1027
Agbosomé	British Jurisdiction over Seaboard	2 Dec., 1879	91	—	393
Aggravie	Cession to Great Britain	12 Aug., 1866	—	—	396
Agotime Tribe	(West Togo District). German Protectorate	Dec., 1887	127	—	628
Agoué	Franco-German Boundary	24 Dec., 1885	78	2	295
,, ,,	,, ,,	1 Feb., 1887	79	—	297
Ahwoonah	Treaty. Great Britain	30 Nov., 1868	91	—	389
Ahy Lagoon	See Tendo River.				
Ajarra	River. Free Navigation	10 Aug., 1889	110	4, § 1	562
,, ,,	Creek. British and French Boundary	10 Aug., 1889	110	4, § 1	561
Aka	River. British and German Boundary. Gold Coast	1 July, 1890	129	4 Annex 2	567 647
Akaba	Fort. Outside Egyptian Territory	8 April, 1892	67	—	266
Akoonfee	Allegiance to Great Britain	Nov., 1884	—	—	388
Akropong	Treaty. Great Britain	30 Mar., 1850	91	—	388
Albatross Rock	British. Letters Patent	27 Feb., 1867	89	—	344
Albert Lake	Great Britain and Congo Free State. Boundary	12 May, 1894	App. 16	2	1009
Albert Edward	Lake and Lake Tanganika. Territory between	12 May, 1894	App. 16	2	1010
,,	,, ,,	22 June, 1894	App. 18	—	1017
Albreda	Fort James. Guarantee to England	3 Sept., 1783	—	—	365
,,	Boundary. French Factory	19 June, 1826	90	—	369
,,	Territory near. Cession to Great Britain	18 Nov., 1850	90	—	374—376

ALPHABETICAL INDEX.

Name of Country, Place, &c.	Subject.	Date of Treaty or other Document.	No. of Doc.	Art.	Page.
Albreda	French Vessels Trading to	15 June, 1826	90	2	368
,,	Cession to Great Britain (except French portion)	5 Jan., 1827	90	Note	368
,,	,, ,,	7 Mar., 1857	106	2	545
,,	French Rights	15 June, 1826	90	—	368
,,	,,	7 Mar., 1857	106	3, 4	545 546
Alcatras	Islands. (Off Senegal). French Occupation	30 Nov., 1887	Note	—	286
Alcoholic Liquors	See Spirituous Liquors.				
Algeria	French Occupation	1830—1844	75	—	287
,,	And Morocco. Boundary	10 Sept., 1844	170	5	802
,,	,,	18 Mar., 1845	171	1, 2, 3	803
,,	Non-introduction of Arms into Tunis	12 May, 1881	190	9	910
Alhucemas	Moors on Frontier	6 May, 1845	185	2	893
,,	,, ,,	24 Aug., 1859	186	6	898
,,	,, ,,	26 April, 1860	187	5, 6	899
,,	,, ,,	5 Mar., 1894	Note	—	902
Allada	See Dahomey.				
Amadib	Egyptian Evacuation	3 June, 1884	67	—	263
Amaponda	(Pondoland). See Great Britain (Cape Colony).				
Amaquatis	British Protection	10 Dec., 1875	H.T.	xv	860
Amatonga	See Great Britain (Tongaland).				
Ambado	Cession by Tajourah to France	14 Dec., 1884	73	—	277
Ambas Bay	British Sovereignty	19 July, 1884	84	—	321
,,	British Possession	29 Apr., 1885	119	Note	597 599
,,	German Protectorate	28 Mar., 1887	—	Note	600
Amphila	To Ras Dumeira. Danakil Coast. Italian Possession	9 Dec., 1888	12	3	11
Andara	Village. German. Boundary, Germany and Portugal	30 Dec., 1866	85	1	323
Angola	Portuguese Province		Note	—	726
,, ,,	And Mozambique. Territories between. French Recognition of Portuguese Rights	12 May, 1886	80	4	300
,,	,, Conditional recognition of, ditto, by Germany	30 Dec., 1886	85	3	324
,,	,, British Protest against Portuguese Claims	13 Aug., 1887	85	—	325

ALPHABETICAL INDEX.

Name of Country, Place, &c.	Subject.	Date of Treaty or other Document.	No. of Doc.	Art.	Page
Angra Pequeña	See Ichaboe and Penguin Islands, and Great Britain (Cape Colony).				
Anjouan	(Johanna). See France and Comoro Islands.				
Ankara	(Madagascar). Cession to France	5 Mar., 1841	168	—	795
,,	,, Nossi-Mitsion	1 June, 1841	168	—	795
Ankoli	Boundary. Uganda	18 June, 1894	App. 17	—	1016
Annobon	Island. Cession to Spain	1 Mar., 1778	181	13	885
,,	See also Fernando Po.				
Anum	(Gold Coast). Fealty to British Crown	3 July, 1888	01	—	309
Aowin	(Gold Coast). British Possessions	26 June, 1891	114	—	574
Appa (or Appah)	British Protectorate. British Flag hoisted	15 Mar., 1884	93	—	417
,, ,,	British Possession of Capital recognised..	10 Aug., 1889	110	4, § 1	561
,, ,,	Protection to Inhabitants. French Portion	10 Aug., 1889	110 Note	4, § 3 —	562 533
Appolonia	British Rights	13, 14 Feb., 1874	91	—	390
Aquamoo	(Gold Coast). Cession to Great Britain	27 July, 1886	01	—	396
,,	British Protection	9 May, 1887	01	—	307
,,	In British Sphere..	Dec., 1887	127	—	628
Arbitration	British. Differences. Ethiopia and Egypt	3 June, 1884	3	6	4
,,	Disagreements. Treaty Powers Berlin Act	26 Feb., 1885	17	12	30
,,	Differences. Great Britain and Portugal	11 June, 1891	150	11, 13	737–798
Archibong's	(Or Arsibon's) Village. Rio del Rey Boundary	14 Apr., 1883	129	3	621
Arguin	Fort. Cession to France..	3 Sept., 1783	103	9	539
Arms and Ammunition	Non-introduction into Algeria through Tunis	12 May, 1881	190	9	910
,,	"Berlin Act." Transport. Congo and Niger	26 Feb., 1885	17	25—33	39—43
,,	Zanzibar and British East Africa Company	9 Oct., 1888	26	4	130
,,	Importation prohibited. Zone defined. "Brussels Act"	2 July, 1890	18	8—13	56—60
,,	Tariff. Eastern Zone. Congo Basin	22 Dec., 1890	19	2	90

ALPHABETICAL INDEX.

Name of Country, Place, &c.	Subject.	Date of Treaty or other Document.	No. of Doc.	Art.	Page.
Arms and Ammunition	British South Africa Company	30 July, 1891	Note	—	186
,,	Import Duties. Zanzibar.	22 June, 1892	App. 11	—	993
,,	Imports. Benadir Ports..	12 Aug., 1892	203	4	953
,,	See also Separate Countries.				
Aroangwa	River. British and Portuguese Spheres of Influence	11 June, 1891	147	1, 2	690
,,	,,	May--June, 1893	151	10	746
Artigere	See Atijere.				
Ashantee	Treaty. Great Britain	27 Apr., 1831	90	—	388
,,	,,	13—14 Feb., 1874	91	—	390
,,	British and French political action..	10 Aug., 1889	110	3, § 1	559
Assab	(Danakils). Cession to Italy of part of Ablis	15 Mar., 1883	5	—	5
,,	,, Intercourse with Shoa. Protection of Natives	21 May, 1883	6	9—11	6
,,	,, Road to Mt. Musalli..	7 July, 1887	8	2	8
,,	,, Assab-Aussa-Shoa Road	10 Aug., 1887	9	2	9
,,	,,	9 Dec., 1888	12	2	11
,,	Spanish Naval Station	Dec., 1887	183	—	888
Assinee	French Customs Tariff	10 Aug., 1889	110	3, § 3	561
,,	British and French Boundary..	10 Aug., 1889	110	3, § 1 Annex 2	560 567
Atbara..	River. British and Italian Spheres of Influence	15 Apr., 1891	136	1, 2	668
Atijere..	Island (Lagos). British Protectorate	24 Oct., 1885	92	—	422
,,	Wharf. Jurisdiction. Itebu, &c.	9 Oct., 1888	92	—	431
Aubad..	Island. Ceded to Great Britain	3 Sept., 1840	178	—	832
,,	British Protectorate	20 July, 1887	178	—	834
Aussa	(Danakils). Cession of part to Italy	15 Mar., 1883	5	5	5
,,	,, Road. Assab to Mt. Musalli	7 July, 1887	8	—	8
,,	,, Assab-Aussa-Shoa Road	10 Aug., 1887	9	—	9
,,	,,	9 Dec., 1888	12	2	11
,,	,, Italian Administration of Justice..	10 Aug., 1887	9	3	9
,,	,, Italian Protectorate and Sovereignty over Danakil Coast	9 Dec., 1888	12	2, 3	11

ALPHABETICAL INDEX.

Name of Country, Place, &c.	Subject.	Date of Treaty or other Document.	No. of Doc.	Art.	Page.
Aussa ..	(Danakils). Exclusion of Foreigners ..	9 Dec., 1888	12	5	11
,, ..	,, Italian Protectorate over Aussa ..	6 Dec., 1889	16	—	18
Austria-Hungary	And Congo. Recognition of Association ..	24 Dec., 1884	40	Note	194 106
Avrekete ..	Annexed to France ..	3 Dec., 1892	65	—	248
Awoonah ..	British Jurisdiction ..	22 June, 1874	91	3	391
Axim ..	See Great Britain (Gold Coast) ..	23 Aug., 1886	89	—	356
Ayliff, Mt. ..	Pondoland. See also Great Britain (Cape Colony) ..	29 July, 1887	89	—	361
Azaphe ..	River to Cape Bellone (Madagascar). French Occupation	4 Nov., 1818	168	—	795
Bab ..	Island. Cession to Great Britain ..	27 Aug., 1840	178	—	832
,, ..	,, Included in French Sphere ..	2—9 Feb., 1888	App. 3	—	976
Bacca Loco ..	Cession to Great Britain..	12 Dec., 1825	100	—	494
Badagry ..	(Lagos). Cession to Great Britain ..	7 July, 1863	92	—	414
,, ..	,, ,,	5 May, 1886	92	—	424
Bafing ..	River. Soudan ..	—	—	—	285
Bagamoyo ..	Zanzibar Possession ..	9 June, 1886	120	5	606
,, ..	German Possession ..	27—28 Oct., 1890	Note	—	650
Bageida ..	Harbour. German Protectorate ..	15 Oct., 1884	84	—	320
Bago ..	Cession to Great Britain Isles de Los ..	6 July, 1818	100	—	485
Bagroo..	Cession to Great Britain..	9 Nov., 1861	100	—	509
Bahr el Ghazal	Excluded from German Sphere ..	15 Nov., 1893	133	4	660
Bakassey or Backasay	Peninsula. Rio del Rey..	14 April, 1893	131	3	655
Bamaquillad ..	Liberian possession ..	8 Dec., 1892	164	1, § 3	784
Bamou Island	(Stanley Pool). French Possession ..	5 Feb., 1895*	App. 31	—	1061
Banana ..	Islands. Cession to Great Britain ..	25 May, 1819	100	—	486
,, ..	,, ,,	21 July, 1820	100	—	487
,, ..	,, ,,	20 Oct., 1820	100	—	487
Bance ..	Islands. (Sierra Leone). Cession to Great Britain ..	10 July, 1807	100	—	485
,, ..	,, ,,	2 Aug., 1824	100	—	489

* Treaty not yet ratified (1 Feb., 1896).

ALPHABETICAL INDEX.

Name of Country, Place, &c.	Subject.	Date of Treaty or other Document.	No. of Doc.	Art.	Page.
Bangweolo	Lake. Boundary. Congo State	1 Aug., 1885	42	—	199
,,	,, ,, rectified	12 May, 1894	App. 16	1	1008
Banjola	Islands. See St. Mary Island.				
Baraka	To Red Sea. Frontier. Italy and Egypt	25 June, 1895	App. 5	—	1072
Barolongs	See Great Britain (Bechuanaland).				
Barotse	In British Sphere..	11 June, 1891	150	4	733 734
,,	,,	May—June, 1893	151	10	746
Barra	Cession of Bacco Loco to Great Britain	12 Dec., 1825	100	—	494
Barruwa	(Lake Chad). French influence	5 Aug., 1890	113	2	572
Bastards	Or Griqua country, annexed to Bechuanaland..	8 May, 1891	88	—	336
Basutoland	Notes on ..	1843—1891	87	—	329
,,	Independence recognized.	23 Feb., 1854	87	—	329
,,	British territory ..	12 Mar., 1868	87	—	330
,,	Annexed to Cape Colony. Boundaries..	11 Aug., 1871	87	—	330
,,	Disannexed from do.	No. 34. 1883	87	—	331
,,	Direct British authority over..	2 Feb., 1884	87	—	332
Batanga	Little. German Protectorate	15 Oct., 1884	84	—	321
Batelling	(Central Kiang). British Sovereignty	14 Oct., 1868	90	—	383
Bathurst	French residents. Consular Agent	7 Mar., 1857	106	3	545
Batlapings	See Great Britain (Bechuanaland).				
Bayol Islands	Slave Coast. Franco-German Boundary.	1 Feb., 1887	79	—	297
Bechuanaland	(British) and the South African Republic. Boundary ..	17 Oct., 1871	179	—	840
,,	,, ,, ,,	3 Aug., 1881	179	—	845
,,	,, ,, ,,	27 Feb., 1884	179	—	851
,,	Country of the Batlapings ..	3 May, 1884	88	—	334
,,	Country of the Barolongs	22 May, 1884	88	—	334
,,	Bechuanaland and the Kalahari. British Jurisdiction	27 Jan., 1885	88	—	334
,,	British Protectorate. Boundaries..	30 Sept., 1885	88	—	335
,,	Government. British Bechuanaland..	29 Sept., 1885	88	—	334
,,	Territories to north of	29 Oct., 1889	88	—	335
,,	,, ,,	30 June, 1890	88	—	336

ALPHABETICAL INDEX.

Name of Country, Place, &c.	Subject.	Date of Treaty or other Document.	No. of Doc.	Art.	Page.
Bechuanaland	British and German spheres	1 July, 1890	129	3	646
,,	Annexation. Bastards or Griqua country	8 May, 1891	88	—	336
,,	British Jurisdiction. Tati, &c.	27 June, 1891	App. 9	—	990
Bechuanaland (British), Bechuanaland, and the Kalahari	Notes on.. British Jurisdiction	1884—1893 27 June, 1891	88 App. 9	— —	334 990
Beilul ..	Italian Administration	10 Aug., 1887	9	3	9
Belgium	And Congo. Recognition of Association	23 Feb., 1885	41	—	196
,,	See also Congo and Belgium.				
,,	And Zanzibar. Consular jurisdiction..	30 May, 1885	194	—	926
,,	Award. Customs and administration. Isle of Lamu	17 Aug., 1889	128	—	630
,,	And France. French right of pre-emption. Congo State	5 Feb., 1895*	App. 30	—	1059
,,	,, Stanley Pool	5 Feb., 1895*	App. 31	—	1061
Belley ..	(Sherbro). Cession to Great Britain	9 Nov., 1861	100	—	509
Benadir Ports	(Brava, Meurka, Magadisho, and Warsheikh). Limits. Zanzibar sovereignty	9 June, 1886 Oct.—Nov., 1886	120 123	8 1	607 616 619
,,	,, ,,	3, 4 Dec., 1886	153	1	755 758
,,	Eventual occupation by Italy	15 Jan., 1889	27	--	137
,,	Eventual transfer by British East Africa Company to Italy ..	3 Aug., 1889	27	—	137
,,	Leased by Zanzibar to British East Africa Company	31 Aug., 1889	154	1	760
,,	Transfer. British East Africa Company to Italy, subject to approval of Zanzibar..	18 Nov., 1889	28	—	142
,,	Concession. Zanzibar to British East Africa Company	4 Mar., 1890	30	—	148

* Treaty not yet ratified (1 Feb., 1896).

ALPHABETICAL INDEX.

Name of Country, Place, &c.	Subject.	Date of Treaty or other Document.	No. of Doc.	Art.	Page.
Benadir Ports	Zanzibar consent to transfer by British East Africa Company to Italy	5 Mar., 1890	31	—	149
,,	Administration by British East Africa Company transferred to Italy	8 April, 1890	202	—	949
,,	Excepted from British Protectorate of Zanzibar	4 Nov., 1890	156	--	706
,,	Concession of 4th Mar., 1890, modified . .	5 Mar., 1891	31	—	150
,,	Import Duties	22 June, 1892	App. 11	—	993
,,	Administration conceded by Zanzibar to Italy	12 Aug., 1892	203	—	950
,,	,, ,, ,, See also Zanzibar Dominions.	15 May, 1893	204	—	958
Bender Ziadeh	See Bunder Ziadah.				
Bendoo.. . .	(Sherbro). Cession of portion to Great Britain	9 Nov., 1861	100	—	508
Benin River . .	British Protectorate, as far as to Odi. Coast line	5 Feb., 1886	92	—	422
,,	British Sovereignty and Protection . . . . See also Niger Coast Protectorate.	5 Feb., 1886	92	—	422
Bennah . .	And Tambakka (Sierra Leone). Anglo-French Boundary . .	10 Aug., 1889	100	—	526
Benué	River. Niger Districts. British and German Spheres. Gulf of Guinea	July—Aug., 1886	122	—	612
,, ...	,, and Lake Chad. Transit Dues, &c. . .	1 July, 1890	129	5	647
,, ...	,, and Lake Chad. Anglo - German Spheres	15 Nov., 1893	133	2	659
Berbera . .	Free Port	7 Sept., 1877	67	1	261
Berim River . .	(Gold Coast). Insuaim Ferry	2 June, 1888	91	—	398
Berlin Act . .	Trade. Civilization. Rivers Congo, Niger, &c. Slave Trade. Occupation of African Coasts	26 Feb., 1885	17	—	20
,,	Ratifications. Do. . .	—	—	—	45
,,	Accessions. Liberia, Persia, Zanzibar	—	—	Note	48

ALPHABETICAL INDEX.

Name of Country, Place, &c.	Subject.	Date of Treaty or other Document.	No. of Doc.	Art.	Page.
Berlin Act	Application to British and German Spheres within Free Zone	1 July, 1890	129	8	648
Bertiri	Tribes. British and Italian Spheres	5 May, 1894	136*	1	670
Betente	Cession to Great Britain	25 July, 1887	90	—	381
Bia Kabouba	Anglo-French Spheres	2—9 Feb., 1888	App. 3	1	976
Bight of Biafra	Franco-German Limits. See also Niger Coast Protectorate.	24 Dec, 1885	78	1	293
Bimbia	Cameroons. German Protectorate	15 Oct., 1884	84	—	320
Binue	See Benué.				
Bizerta	Non-annexation by France	20 May, 1881	108	—	552
"	British Right to use Port	20 May, 1881	108	—	553
Blue Nile	To Juba River. Anglo-Italian Spheres	24 Mar., 1891	135	—	665
	To Ras Kasar, Anglo-Italian Spheres	15 Apr., 1891	136	—	667
Bogos	Restoration to Ethiopia	3 June, 1884	3	2	3
Bompey Island	Limits	4 July, 1849	100	—	501
Bomvanaland	See Great Britain (Cape Colony).				
Bonkia	(Quiah). Sierra Leone	20 Jan., 1872	100	—	513
Bonny	See Niger Coast Protectorate.				
Borgu	(Boussa). Royal Niger Company. British Protection, &c.	20 Jan., 1890	App. 4	—	981
Boundaries	See Separate Countries.				
Boussa	See Borgu.				
Brava	Zanzibar Limits. See also Benadir Ports.	7 June, 1886	120	8	607
Brekama	British Sovereignty and Protection	29 May, 1827	90	—	369
British Bechuanaland	See Great Britain (Bechuanaland).				
British Central Africa	Protectorate	22 Feb., 1893	173	—	811
"	Administration by British South Africa Company	24 Nov., 1894	App. 24	—	1025
British Colonies	In Africa. See Great Britain List facing page 326.				
British East Africa Company	Concession. Sultan of Zanzibar to British East Africa Association. Mrima. Wanga to Kipini. (See also Concession, 9th October, 1888)	24 May, 1887	24	—	110
" "	Royal Charter	3 Sept., 1888	25	—	118

ALPHABETICAL INDEX.

Name of Country, Place, &c.	Subject.	Date of Treaty or other Document.	No. of Doc.	Art.	Page.
British East Africa Company	Concession. Sultan of Zanzibar to British East Africa Company. Extension of Privileges. Mrima. Wanga to Kipini and Islands. (See also Concession, 4th March, 1890) ..	9 Oct., 1888	26	—	125
,, ,,	Agreement. British East Africa Company and Italy. Concession of Kismayu, Brava, Meurka, Magadisho, and Warsheikh to be made by Company to Italy when conceded by Zanzibar to Company ..	3 Aug., 1889	27	—	137
,, ,,	Deed. Transfer by British East Africa Company to Italy of Company's Rights over Brava, Meurka, Magadisho, and Warsheikh. Joint Occupation of Kismayu	18 Nov., 1889	28	—	142
,, ,,	Agreement. British East Africa Company and Zanzibar. Customs Revenues. Wanga to Kipini	21 Dec., 1889	29	—	146
,, ,,	Concession. Sultan of Zanzibar to British East Africa Company. Kipini to Mruti. Lamu. Manda. Patta. Kwyho. Benadir Ports:—Kismayu, Brava, Meurka Magadisho, and Warsheikh. (Amended by Agreement of 5th March, 1891) ..	4 Mar., 1890	30	—	148

ALPHABETICAL INDEX.

Name of Country, Place, &c.	Subject.	Date of Treaty or other Document.	No. of Doc.	Art.	Page.
British East Africa Company	Suppl. Agreement. Modification by Sultan of Zanzibar of Concession to British East Africa Company of 4th March, 1890. Wanga to Kipini; Lamu, Manda, Patta, and Kismayu conceded to Company "in perpetuity." Benadir Ports:—Brava, Meurka, Maradisho, Warsheikh, and Mruti undisturbed	5 Mar., 1891	31	—	150
,, ,,	And British Government Administration of Witu	5 Mar., 1891	32	—	153
,, ,,	,, ,, ,,	18 Mar., 1891	33	--	157
,, ,,	,, ,, Notification ditto and Coast. Kipini to Kismayu	20 Mar., 1891	34	—	160
,, ,,	List of Treaties. British East Africa Company and Native Chiefs	1887—1891	35	—	164
,, ,,	Notes. On Boundaries of British Sphere of Influence on East Coast of Africa	1887—1891	36	—	170
,, ,,	Withdrawal. North of the Tana. See also Uganda. Witu.	31 July, 1893	80	—	327
,, ,,	Zanzibar Administration. British Protectorate	31 July, 1893	86	--	328
,, ,,	And Italy. Eventual Concession of Kismayu and Benadir Ports to, by Company	3 Aug., 1889	27	—	137
,, ,,	,, ,, ,, Transfer by Company to Italy. Benadir Ports. Joint Occupation of Kismayu	18 Nov., 1889	28	—	142
,, ,,	,, ,, ,, Consent of Sultan of Zanzibar to ditto..	5 Mar., 1890	30	—	149
,, ,,	,, ,, ,, Notification of Transfer..	8 Apr., 1890	202	—	949
,, ,,	,, ,, ,, Zanzibar to Italy. Benadir Ports	12 Aug., 1892	203	—	950
,, ,,	And Native Chiefs. Treaties	June, 1887	35	—	164

ALPHABETICAL INDEX.

Name of Country, Place, &c.	Subject.	Date of Treaty or other Document.	No. of Doc.	Art.	Page.
British East Africa Company	And Native Chiefs. Treaties	1887—1891	35	—	184
,, ,,	And Zanzibar. Concession. Wanga to Kipini, &c... ..	24 May, 1887	24	—	110
,, ,,	,, ,, ,, Including Extension and Islands.. ..	9 Oct., 1888	26	—	125
,, ,,	And Zanzibar. Administration of Territory on Mainland and Islands ..	31 Aug., 1889	154	—	760
,, ,,	,, ,, ,, Customs. Wanga to Kipini	21 Dec., 1889	29	—	146
,, ,,	,, ,, ,, Kipini to Mruti and Benadir Ports ..	4 Mar., 1890	30	—	148
,, ,,	,, ,, ,, Concession. 4 March, 1890, Modified. Granted in perpetuity	5 Mar., 1891	31	—	150
,, ,,	Import Duties. Zanzibar Ports	22 June, 1892	App. 11	—	993
British East Africa	Arrangements. Great Britain and Germany. Spheres of Influence	Oct.—Nov., 1886	123	—	615
,, ,,	,, ,, ,,	1 July, 1890	129	—	642
,, ,,	,, "British Protectorate over Witu, &c., and Coast from Kipini to Kismayu	19 Nov., 1890	86	—	327
,, ,,	,, Great Britain and Italy. Spheres of Influence. River Juba to Blue Nile ..	24 Mar., 1891	135	—	665
,, ,,	,, Administration by Zanzibar of Protectorate North of the Tana, with exceptions	31 July, 1893	160	—	770
,, ,,	British Protectorate. Uganda	27 Aug., 1894	App. 22	—	1023
,, ,,	Late Possessions of British East Africa Company	15 June, 1895	App. 36	—	1069

ALPHABETICAL INDEX.

Name of Country, Place, &c.	Subject.	Date of Treaty or other Document.	No. of Doc.	Art.	Page.
British East Africa	Territory transferred. British Flag hoisted ... See also Uganda. Witu.	1 July, 1895	App. 36	—	1069
British Protectorates	See Protectorates.				
British South Africa Company	Royal Charter	29 Oct., 1889	37	—	174
,, ,,	Notes on Nyasaland, &c...	1890, 1891	38	—	183
,, ,,	Treaties. Native Chiefs. Shiré Highlands	1889—1893	39	—	187
,, ,,	Operations north of the Zambesi	Feb., 1891	App. 8	—	987
,, ,,	Administration. British Central Africa	24 Nov., 1894	App. 24	—	1025
,, ,,	Conditional confirmation of Treaties	24 Nov., 1894	App. 24	—	1025
British Spheres of Influence	East Coast of Africa. Notes. See also Protectorates. Separate Countries and Spheres of Influence.	1887—1891	36	—	170
Broussa	(Gold Coast). British Possession	26 June, 1891	91	—	401
Brussels Act*	African Slave Trade	2 July, 1890	18	—	48
,,	Central Office	2 July, 1890	18	82	82
,,	French exception to certain Articles	2 Jan., 1892	21	—	98
,,	Application of, to certain French Possessions.	2 Jan., 1892	21	—	99
,,	Ratifications. Various Powers	2 July, 1891	20	—	92
,,	,, ,,	2 Jan., 1892	21	—	97
,,	,, United States	2 Feb., 1892	22	—	102
,,	,, Portugal	30 Mar., 1892	23	—	104
,,	Accession. Ethiopia	16 Sept., 1890	18	Note	48
,,	,, Liberia	7—25 Aug., 1892	18	Note	48
,,	,, Orange Free State	24 Nov.—31 Dec., 1894	18	Note	48
,,	,, Persia	3 July, 1890	18	Note	48
Buem	(Gold Coast). Fealty to British Crown	3 July, 1888	91	—	399
Buffalo River	See South African Republic.				
Bulama Island	Cession to Great Britain	3 Aug., 1792	100	—	498
,,	,, ,, ,,	24 June, 1827	100	—	498

* Came into force April 2, 1894.

ALPHABETICAL INDEX.

Name of Country, Place, &c.	Subject.	Date of Treaty or other Document.	No. of Doc.	Art.	Page.
Bulama Island	Dispute. Great Britain and Portugal referred to Arbitration	13 Jan., 1869	139	—	684
,,	Awarded to Portugal	21 April, 1870	140	—	688
,,	Portuguese occupation	1 Oct., 1870	141	—	691
Bulhar	Free Port	7 Sept., 1877	67	1	261
Bullom	And Shebar (Sierra Leone). British Sovereign rights. See also Kafu Bulloms and North Bulloms.	18 Nov., 1882	100	—	520
Bunder Ziadeh	To Ras Jiburti. Somali Coast. British Protectorate	20 July, 1887	178	...	834
,,	French recognition. Do...	2—9 Feb., 1888	App. 3	—	976
Burruwa	See Barruwa.				
Busi	River. Transit. Persons and Goods	11 June, 1891	150	12	737
Cabinda	Portuguese Claim. Boundary. Congo and Portugal	28 July, 1817	138	—	683
,,	,, ,,	14 Feb., 1885	58	3	232
,,	,, ,,	25 May, 1891	60	2	237
Cabinda Bay	Congo. Limits	1 Aug., 1885	42	—	198
Cabo Lombo	Portuguese Boundary. Congo State	14 Feb., 1885	58	—	232
,,	Congo. Limits	1 Aug., 1885	42	—	198
Cabompo River	British and Portuguese Spheres	May—June, 1893	151	5	744
Cajet River	Guinea Coast	12 May, 1886	80	1	299
Calabar (Old and New)	See Niger Coast Protectorate.				
Cameroons	German Protectorate	12 July, 1884	84	—	320
,,	,,	15 Oct., 1884	84	—	320
,,	British and German Spheres	April—June, 1885	119	—	614—596
,,	Franco-German Boundary	24 Dec., 1885	78	—	293
,,	,, ,,	4 Feb., 1894	App. 13	—	998
Campo River	See also Guinea (Gulf of). Boundary. France and Germany	24 Dec., 1885	78	1	294
Cape Bellone	(Madagascar) to River Azappe, French Occupation	4 Nov., 1818	168	—	795
Cape Blanco	To Cape Bojador. Spanish Protectorate	9 Jan., 1885	180	—	883
Cape Bojador	See Cape Blanco.				

ALPHABETICAL INDEX.

Name of Country, Place, &c.	Subject.	Date of Treaty or other Document.	No. of Doc.	Art.	Page.
Cape Colony ..	Cession by Holland to Great Britain ..	13 Aug., 1814	137	—	672
,, ..	Eastern Boundary ..	10 May, 1835	89	—	341
,, ..	,, ,, ..	16 June, 1835	89	—	341
,, ..	North-Eastern Boundary .	14 Oct., 1835	89	—	342
,, ..	Annexation of Congo, Gaika, and T'Slambie Territories ..	17 Sept., 1835	80	—	342
,, ..	Renunciation of British Authority over ditto, ditto..	5 Dec., 1836	89	—	342
,, ..	New Boundaries	5 Dec., 1836	89	—	342
,, ..	,, ,, ..	2—30 Jan., 1845	80	—	343
,, ..	Incorporation of British Kaffraria with Cape Colony ..	17 Mar., 1865	89	—	343
,, ..	Ichaboe and Penguin Islands	1861—1886	89	—	345
,, ..	Fingoland. Idutywa Reserve and Nomansland. Annexation to Cape Colony ..	12 June, 1876	89	—	343
,, ..	Pondoland. Unzimkulu, and Unzimabu Rivers. Cession to Cape Colony ..	17 July, 1878	89	—	346
,, ..	,, British Protectorate over Coast ..	5 Jan., 1885	80	—	351
,, ..	,, St. John's River. Port and Tidal Estuary. Annexation to Cape Colony	4 Sept., 1878	89	—	347
,, ..	,, ,, ,,	22 July, 1884	89	—	351
,, ..	,, Boundaries of Amaponda	7 Oct.— 23 Nov., 1884	89	—	346
,, ..	,, Xesibeland. Mt. Ayliff. Annexed to Cape Colony ..	23 Aug., 1886	89	—	351
,, ..	,, Rode Valley, St. John's River, Mt. Ayliff and Mt. Frere Districts. Cessions to Cape Colony ..	9 Dec., 1886	89	—	354
,, ..	,, Road, Eastern Pondoland to south of St. John's River .	9 Dec., 1886	89	—	355
,, ..	,, Rode Valley. Mt. Frere. Griqualand East. Annexation to Cape Colony ..	29 July, 1887	89	—	357
,, ..	,, British Sovereignty. Pondoland	20 Mar., 1894	89	—	358

ALPHABETICAL INDEX.

Name of Country, Place, &c.	Subject.	Date of Treaty or other Document.	No. of Doc.	Art.	Page.
Cape Colony ..	Walfisch Bay. See Walfisch Bay.				
Cape of Good Hope	Notes on	1652—1894	89	—	341
,, ..	See also Great Britain (Cape Colony).				
Cape Frio ..	To Orange River. German Protectorate (except Walfisch Bay)..	15 Oct., 1884	83	—	318
Cape Juby ..	Cession	19 April, 1879	—	—	105
,, ..	Purchase by Morocco of N. W. Africa Co.'s Territory	13 Mar., 1895	App. 33	—	1064
Cape Roxo ..	Guinea Coast	12 May, 1886	80	1	298
Cape Spartel ..	Lighthouse	31 May, 1865	172	—	808
Catack Island	Guinea Coast. Portuguese	12 May, 1886	80	1	299
Catima Rápids	Upper Zambesi. Boundary. Germany and Portugal	30 Dec., 1886	85	1	323
Cavalla ..	Submission to Liberia ..	10 Nov., 1893	164	—	786
Cavalla River .	To San Pedro River. Liberian Claim ..	1891—1892	{ 115 164	— —	576 783
,, ..	Boundary. France and Liberia	8 Dec., 1892	164	1	784
,, ..	Free Navigation	8 Dec., 1892	164	2	784
,, ..	Liberian Sovereignty. West of Grain Coast	8 Dec., 1892	164	3	785
,, ..	To Lahou. French Sovereignty and Protection	1891—1892	115	—	576— 518
Ceded Mile ..	(Gambia). See Great Britain (Gambia).				
Central Angoniland	Mining Rights	24 Nov., 1894	App. 24	—	1025
Central Kiang	(Gambia.) British Sovereignty	19 Sept., 1887	90	—	383
,,	(Batelling) Sovereignty ..	14 Oct., 1888	90	—	383
Ceuta	Limits	7 Oct., 1844	184	—.	891
,, ...	,, Extended ..	26 April, 1860	187	2	897
,, ...	Frontier. Larache ..	6 May, 1845	185	—	893
,, ...	Neutral Ground	26 April, 1860	187	36	897 898
Chad	Lake and River Benue. Transit Dues. Treaties	1 July, 1890	129	5	647
,, ...	(Barruwa). French Influence	5 Aug., 1890	113	2	572
,, ...	Anglo-German Boundary.	15 Nov., 1893	133	2	659
,, ...	And French Congo. German Limits.. ..	4 Feb., 1894	App. 13	—	998

ALPHABETICAL INDEX.

Name of Country, Place, &c.	Subject.	Date of Treaty or other Document.	No. of Doc.	Art.	Page.
Chagga	Treaties. German Colonization Society	1885	82	—	309
,,	British and German Spheres	Oct.—Nov., 1886	123	3	617—620
,,	,, ,,	3, 4 Dec., 1886	153	4	756 759
,,	,, ,, See also Kilimanjaro District.	1 July, 1890	129	1	642
Chah	(Sherbro). Cession of portion to Great Britain	9 Nov., 1861	100	—	508
Chaki Chaki	See Pemba Island.				
Chamba	Point. Guinea Coast	12 May, 1886	80	3	300
Charters	German Colonization Society	18 Feb., 1885	81	—	303
,,	National Africa Company	10 July, 1886	96	—	446
,,	Royal Niger Company	10 July, 1886	96	—	446
,,	Imperial British East Africa Company	3 Sept., 1888	25	—	118
,,	British South Africa Company	29 Oct., 1889	37	—	174
Chiloanga	Boundary. Congo	1 Aug., 1885	42	—	198
,,	,, France and Portugal	12 May, 1886	80	3	300
Chilwa	Lake. British and Portuguese Limits	May—June, 1893	151	1	743
Chindé	North. Lease. Land. To a British Company	20 Aug., 1890	148	13	724
Chinta	Lake. British and Portuguese Spheres	11 June, 1891	151	1	743
Chisamulu	Island. Lake Nyasa. British	May—June, 1893	151	4	744
Chobé	River. British and German Spheres	1 July, 1890	129	3, § 2	645
Coba	German recognition. French rights	24 Dec., 1885	78	3	296
Combo	(Gambia). Cession to Great Britain	13 July, 1840	90	—	373
,,	,, ,, ,,	26 Dec., 1850	90	—	378
,,	,, British Combo	24 May, 1853	90	—	380
,,	,, Annexation. Foreign Combo	April, 1894	—	—	384
Comoro	(Grand Comoro). French Protectorate	26 June, 1886	77	—	292
,,	,, Application of "Brussels Act" to. See also Johanna. Mohilla.	2 Jan., 1892	121	—	99

ALPHABETICAL INDEX.

Name of Country, Place, &c.	Subject.	Date of Treaty or other Document.	No. of Doc.	Art.	Page.
Congo	"International Association." Name changed to "Independent State of the Congo"	—	Note	—	195
Congo Basin ..	Freedom of Trade. Berlin Act	26 Feb., 1885	17	1 10—12	22—24 29—30
,, ..	Neutrality of Territories and Territorial Waters	26 Feb., 1885	17	1 10—12	22—29 29—30
,, ..	International Navigation Commission.. ..	26 Feb., 1885	17	8 17	28 34
,, ..	Free Navigation during War..	26 Feb., 1885	17	25	38
,, ..	British and German Spheres	1 July, 1890	129	1, § 2	643
,, ..	Tariff. Eastern Zone ..	22 Dec., 1890	19	—	90
,, ..	,, Zanzibar	22 June, 1892	App. 11	—	993
,, ..	British and Congo State Spheres. Watersheds between Nile and Congo	12 May, 1894	App. 16	1—2	1008
Congo Free State	And Austria. Recognition of Association..	24 Dec., 1884	40	—	194
,,	And Belgian Expedition Treaties	1882—1884	—	—	200
,,	And Belgium. Recognition of Association..	23 Feb., 1885	41	—	196
,,	Neutrality. Limits ..	1 Aug., 1885	42	—	198
,,	,, States and Territorial Waters ..	26 Feb., 1885	17	10	29
,,	Administrative Districts ..	1 Aug., 1888	43	—	201
,,	,, Eastern Kwango ..	10 June, 1890	44	—	204
,,	Tariff. Eastern Zone. Congo Basin ..	22 Dec., 1890	19	—	90
,,	,, Zanzibar ..	22 June, 1892	App. 11	—	993
,,	And Denmark. Recognition of Association	23 Feb., 1885	45	—	205
,,	And France. French right of Pre-emption	April—May, 1884	46	—	207
,,	,, ,, ,, ..	April, 1887	50	—	215
,,	,, Recognition of Association, Boundaries, &c.	5 Feb., 1885	47	—	209
,,	,, Private Stations and Properties ..	5 Feb., 1885	48	—	212

ALPHABETICAL INDEX.

Name of Country, Place, &c.	Subject.	Date of Treaty or other Document.	No. of Doc.	Art.	Page.
Congo Free State	And France. Manyanga Region..	22 Nov., 1885	49	—	213
,,	,, Oubangi Region ..	29 April, 1887	51	—	217
,,	,, Boundaries	14 Aug., 1894	App. 21	—	1021
,,	And Germany. Recognition of Association..	8 Nov., 1884	52	—	219
,,	And Great Britain. Recognition of Association	16 Dec., 1884	53	—	221
,,	,, Consular Jurisdiction, &c.	16 Dec., 1884	54	—	223
,,	,, Spheres of Influence. Territories leased, &c. ..	12 May, 1894	App. 16	—	1008
,,	Art. 3. Ditto. Withdrawn	22 June, 1894	App. 18	—	1017
,,	And Italy. Recognition of Association	19 Dec., 1884	55	—	227
,,	And Liberia. Commercial Intercourse, &c. ..	15 Dec., 1891	56	—	229
,,	,, Accession to "Brussels Act" ..	25 Aug., 1892	—	Note	48
,,	And Netherlands. Recognition of Association	27 Dec., 1884	57	—	230
,,	And Portugal. Ditto ..	14 Feb., 1885	58	—	232
,,	,, Tariff. Congo. Basin ..	9 Feb., 1891	—	Note	238
,,	,, ,, ,, ..	8 Apr., 1892	—	Note	238
,,	,, Lunda Region ..	25 May, 1891	59	—	234
,,	,, Boundaries. Lower Congo ..	25 May, 1891	60	—	236
,,	,, Ditto. Lunda Region ..	24 Mar., 1894	App. 15	—	1004
,,	And Russia. Recognition of Association ..	5 Feb., 1885	61	—	239
,,	And Spain. Ditto	7 Jan., 1885	62	—	240
,,	And Sweden and Norway. Ditto ..	10 Feb., 1885	63	—	242
,,	And United States. Ditto See also "Berlin Act," "Brussels Act."	22 Apr., 1884	64	...	244
,,	And Lake Nyasa. Transit Dues ..	1 July, 1890	129	8	649
,,	Cession to Belgium ..	9 Jan., 1895*	App. 27	—	1041
,,	Neutrality.. ..	11 Jan., 1895*	App. 28	—	1043
,,	French right of Pre-emption.. ..	5 Feb., 1895	App. 30	—	1059

* Treaty not yet ratified (1 Feb., 1896).

ALPHABETICAL INDEX.

Name of Country, Place, &c.	Subject.	Date of Treaty or other Document.	No. of Doc.	Art.	Page.
Congo, French	Franco-German Boundaries	4 Feb., 1894	—	—	98
Congo ..	Boundary North of German Sphere	12 May, 1894	App. 16	1A	1008
,, ..	Ditto. North of the Zambesi ..	12 May, 1894	App. 16	1A	1008
,, ..	(Kaffraria). Annexation to Cape Colony. See also Great Britain (Cape Colony).	17 Sept., 1835	89	—	342
Congo Region	French and Portuguese Limits	12 May, 1886	80	3	299
,,	British and German Spheres	1 July, 1890	129	1, § 3	643—644
,,	,, Transit Dues. Lake Nyasa	1 July, 1890	129	8	649
Congo River ..	Free Navigation ..	26 Feb., 1885	17	2 4 13—25	25 23 30—39
,, ..	Portuguese Sovereignty .. (Not ratified).	26 Feb., 1884	147	—	713
Congo Watershed	.. .. .. See also Congo Basin.	12 May, 1894	App. 16	1A	1008
Consular Jurisdiction	(British). In Tunis abolished	31 Dec., 1883	—	Note	915
,,	(British). Congo	16 Dec., 1884	54	5—10	224
,,	(British). Abolition. Massowah ..	17 Aug., 1888	134	—	664
,,	(Italian) in Shoa ..	21 May, 1883	6	12, 13	7
,,	(Ditto) in Ethiopia	1 Oct., 1889	14	9	17
,,	In Zanzibar. See Zanzibar. See also Congo and Separate States.				
Conta ..	(Sierra Leone). Cession to Great Britain ..	18 April, 1826	100	—	495
Corisco Bay ..	Spanish Claim to Territory ..	—	180	—	883
Corteemo Island	(Sierra Leone). See Great Britain. (Sierra Leone).				
Creek Town ..	Old Calabar. See Great Britain (Niger).				
Crepee ..	(Peki). Gold Coast. Fealty to British Crown	17 June, 1888 3 July, 1888	91 91	— —	388 399
,, ..	In British Sphere ,, See also Volta.	Dec., 1887	127	—	628
Criby ..	(Batanga). German Protectorate	15 Oct., 1884	84	—	320

ALPHABETICAL INDEX.

Name of Country, Place, &c.	Subject.	Date of Treaty or other Document.	No. of Doc.	Art.	Page.
Cross River	British. See Great Britain (Niger).				
,,	Boundary. Great Britain and Germany. Gulf of Guinea	April—June, 1885	119	—	596
,,	,, ,,	July—Aug., 1886	122	—	612
Cuango	See Kwango.				
Culacalla	River. Congo. Limits	1 Aug., 1885	42	—	198
Cunene	River. See Kunene.				
Customs Dues	Sherbró and Mendi. British right to	18 April, 1826	100	—	514
,,	Wanga to Kipini	24 May, 1887	24	9	116
,,	,, ,,	9 Oct., 1888	25	—	120
,,	,, ,,	5 Mar., 1891	31	—	151
,,	Kipini to Mruti	31 Aug., 1889	154	—	760
,,	,, ,,	21 Dec., 1889	29	—	146
,,	,, Lamu	4 Mar., 1890	30	3, 6	149
,,	,, ,, ,,	31 Aug., 1889	154	—	760
,,	Benadir Ports. Brava, &c. See also Dar-es-Salaam and Pangani.	31 Aug., 1889	154	—	760
Dafûr	Egyptian Province	13 Feb., 1841	67	—	259
,,	Excluded from German Influence	15 Nov., 1893	133	4	660
Dahomey	And Great Britain	1847—1876	65	—	247
,,	And Portugal. Protectorate. Coast	5 Aug., 1885	65	—	253
,,	,, Protectorate withdrawn	21 Jan., 1886	65	—	253
,,	,, ,,	22 Dec., 1887	65	—	253
,,	French Protectorate	3 Dec., 1892	65	—	248
,,	Notes on Treaties with France, &c.	1847—1893	65	—	248
,,	Submission to France	5 Jan., 1894	App. 13	—	998
,,	King of Abomey. French Protection	29 Jan., 1894	App. 13	—	998
,,	King of Allada. Ditto	4 Feb., 1894	App. 13	—	998
,,	French Colony	22 June, 1894	App. 13	—	998
Dakka	River. Boundary. Great Britain and Germany	1 July, 1890	129	4, § 1	647
Damaraland	German Protectorate	16 Aug., 1884	82	—	317
,,	,, Hereros	21 Oct., 1885	83	—	319
,,	,, British and German Spheres	1 July, 1890	129	3	645
Danakils	Italian Protectorate. Coast	15 Mar., 1883	5	8	5
,,	,, ,,	9 Dec., 1888	12	2	11
,,	,, (Aussa)	6 Dec., 1889	16	—	18

ALPHABETICAL INDEX.

Name of Country, Place, &c.	Subject.	Date of Treaty or other Document.	No. of Doc.	Art.	Page.
Danakils	Italian Sovereignty. Ras Amphila to Ras Dumeira	9 Dec., 1888	12	3	11
,,	See also Abyssinia, &c., Assab, Aussa.				
,,	Cession to France. Obock, &c.	11 Mar., 1862	68	—	269
	See also Tajourah.				
Dar-es-Salaam	Zanzibar Possession	9 June, 1886	120	4	606
,,	Customs Leased to German African Company	Oct.—Nov., 1886	123	2	616—619
,,	,, ,,	3, 4 Dec., 1886	153	2	755—758
,,	German Possession	27, 28 Oct., 1890	—	Note	650
Darfur	See Dafûr.				
Darmi	British and Italian Spheres	5 May, 1894	136*	—	670
Dauphin	Fort (Madagascar). French Occupation	1 Aug., 1819	168	—	795
Dchawe River	Boundary. Great Britain and Germany	1 July, 1890	129	4, § 1	647
Debra Bizen	Convent. Ethiopia	2 May, 1889	13	4	13
Deine	River. Anglo-German Boundary. Gold Coast	1 July, 1890	129	4, § 1	607
Dekami	(Lagos). Part of Frah	23 July, 1886	92	—	424
Delagoa Bay	Portuguese Jurisdiction	28 July, 1817	138	2	683
,,	Dispute. Great Britain and Portugal. Reference to Arbitration	25 Sept., 1872	142	—	693
,,	British Case	Sept., 1873	143	—	694
,,	Non-cession. British or Portuguese Sovereign Rights. Disputed Territory	June, 1875	144	—	697
,,	Award. Franco. Dispute. Great Britain and Portugal	24 July, 1875	145	—	701
,,	And Natal. Coast between. Non-acquisition by Germany	April—May, 1885	119	—	597—599
,,	Portuguese Sphere. Extension south of	11 June, 1891	150	3	733
	See also Lorenço Marques.				
Delgado, Cape	To Bay of Lorenço Marquez. Portuguese. See also Delagoa Bay.	28 July, 1817	138	1	683

ALPHABETICAL INDEX.

Name of Country, Place, &c.	Subject.	Date of Treaty or other Document.	No. of Doc.	Art.	Page.
Denham Waters	(Lagos) Kingdom of Frah	23 July, 1886	92	—	425
Denmark	Cession to Great Britain. Possessions. Gold Coast	17 Aug., 1850	66	—	256
,,	And Congo. Recognition of Association	23 Feb., 1885	45	—	205
Dicomo	Island. See Lukomo.				
Diego Saurez	Madagascar. French occupation	17 Dec., 1885	166	15	793
Dixcove	See Great Britain (Gold Coast).				
Djaliba	Or Niger. See Great Britain (Niger).				
East Equatorial Africa	Anglo-German Boundary.	27 Oct.—24 Dec., 1892	130	—	652
,,	,, ,, ,,	25 July, 1893	132	—	656
,,	Provinces. Great Britain and Belgium (Congo)	12 May, 1894	App. 16	—	1006
,,	See also Dafûr, Kalabat and Kordofan.				
Eastern Kwango	District. Congo	10 June, 1890	44	—	204
	See also Lundi.				
Eastern Soudan	Separate Governorship	30 Nov., 1881	67	—	262
	See also Kalabat. Senhit.				
Ebony Mines	South-West Africa. British Rights	15 July, 1886	121	1	608
Efat	See Abyssinia, &c.				
Egbas	French Trade with	10 Aug., 1889	110	4, § 2	562
Egypt	Treaty. Great Britain and Ethiopia	3 June, 1884	3	—	2
,,	Notes on Boundaries	1840—1894	67	—	259
,,	Non-cession. Somali Territory, &c.	7 Sept., 1877	67	2	261
,,	Non-cession of Rights, &c.	2 Aug., 1879	67	—	262
,,	Confines, to River Juba. British and German Spheres	1 July, 1890	129	4, § 2	644
,,	Rights reserved. Kassala, &c.	15 April, 1891	136	2	668
,,	Claims. Basin of Upper Nile	12 May, 1894	App. 16	—	1008
,,	See also Dafûr, Kordofan, and Bahr-el-Ghazal, and Italy.				
Ela Dabaina	Bedouins. Eastern Soudan	30 Nov., 1881	—	67	262
Elephant	Island. See Delagoa Bay.				
Elmina	Ashantee Renunciation	13—14 Feb., 1874	91	—	390

ALPHABETICAL INDEX.

Name of Country, Place, &c.	Subject.	Date of Treaty or other Document.	No. of Doc.	Art.	Page.
Elobey	(Corisco Bay). Spanish claim	—	180	—	883
Emigrant Tembuland	(Transkei). See Great Britain (Cape Colony).				
Epé	See Great Britain (Lagos).				
Ethiopia	Treaty. Sardinia..	10 Feb., 1859	4	—	4
,,	,, Great Britain and Egypt	3 June, 1884	3	—	2
,,	Boundaries. Italy	2 May, 1889	13	3	13
,,	,, ,,	1 Oct., 1889	14	—	15
,,	Foreign Relations. Italy.	2 May, 1889	13	17	14
,,	Conduct of Foreign Affairs. Italy..	12 Oct., 1889	15	—	17
,,	Preferential Treatment. Italy	2 May, 1889	13	18	15
,,	Italian Jurisdiction	1 Oct., 1889	14	9	17
,,	Italian recognition of King Menelek	1 Oct., 1889	14	1	16
,,	Accession to "Brussels Act" See also Abyssinia and Italy. Massowah.	16 Sept., 1890	Note	—	48
Factories	In Africa. Restored to France	30 May, 1814	104	8	540
Falabah	British Possession. North of Sierra Leone	10 Aug., 1889	110	2 Annex 1	559
,,	And Kambia. English Road	10 Aug., 1889	110	—	564
Fatiko	Headstream of the Niger.	26 June, 1891	114	—	574
Fattatenda	Cession to Great Britain..	13 Apr., 1829	90	—	371
Ferighna	(Sierra Leone). Cession to Great Britain	18 Apr., 1826	100	—	495
Fernando Po	Cession to Spain	1 Mar., 1778	181	—	885
,,	,, ,,	24 Oct., 1778	181	—	884
,,	British Recognition of Spanish Rights	28 Oct., 1830	181	—	885
Fingoland	Annexation to Cape Colony	12 June, 1876	89	—	343
—	See also Great Britain (Cape Colony).				
Fire Arms	See Arms and Ammunition.				
Flag	Congo Free State. See Congo.				
,,	Territory leased by Great Britain to Congo State	12 May, 1894	App. 16	2	1009
,,	National Africa Company (Royal Niger Company)	10 July, 1886	96	11	447
,,	British East Africa Company..	3 Sept., 1888	25	15	123
,,	British South Africa Company..	29 Oct., 1889	37	19	179

ALPHABETICAL INDEX.

Name of Country, Place, &c.	Subject.	Date of Treaty or other Document.	No. of Doc.	Art.	Page.
Flag	Fraudulent use of. "Brussels Act"	2 July, 1890	18	25 51—61	64 73-76
„	Use of, and Supervision by Cruisers. "Brussels Act"	2 July, 1890	18	30—33 41	65-66 C9
„	Witu	5 Mar., 1891	32	7	155
„	„	18 Mar., 1891	33	2	157
Fodedougou-Ba	River. Boundary. France and Liberia	8 Dec., 1892	164	1, § 1	784
„	„ Basin. French Possession	8 Dec., 1892	164	1, § 2	784
Fogni	British. Non-cession to any Foreign Power	15 Sept., 1887	90	—	382
Forcades	See Great Britain (Niger Coast Protectorate).				
Fort Charter	British Jurisdiction	27 June, 1891	App. 9	—	990
Fort Salisbury	„ „	27 June, 1891	App. 9	—	990
Fort Victoria	„ „	27 June, 1891	App. 9	—	990
Fouricaria	(Sierra Leone). Boundaries	2 Aug., 1851	100	—	502
Fouta-Djallon	France and Portugal. Limits	12 May, 1889	80	2	299
„	French Route. Mellicourie and French Soudan	10 Aug., 1889	110	Annex 1	564
„	Trade. Open Roads, &c.	21 Jan., 1895	App. 29	—	1054
Frah	Kingdom (Lagos)	23 July, 1886	92	—	424
France, Isle of	See Mauritius.				
France	Colonies, &c., restored	30 May, 1814	104	8	540
„	And Africa. (East Coast.) Treaties, &c.	1862—1884	68—73	—	269—277
„	„ Treaty. Danakils. Cession of Obock to France	11 Mar., 1862	68	—	269
„	„ French Notice. Limits of Obock	25 Dec., 1880	69	—	272
„	„ Treaty. Gobad. Friendship, &c.	9 Apr., 1884	70	—	273
„	„ Treaty. Tajourah. Cession of Tajourah to France. Gubbed-Kharab, &c.	21 Sept., 1884	71	—	274
„	„ Act. Cession by Sultan of Tajourah to France of Gubbed-Kharab, Rasala, and Sagallo	18 Oct., 1884	72	—	276

ALPHABETICAL INDEX.

Name of Country, Place, &c.	Subject.	Date of Treaty or other Document.	No. of Doc.	Art.	Page.
France..	And Africa. (East Coast.) Act. Cession by Sultan of Tajourah to France of Territory between Adaeli and Ambado ..	14 Dec., 1894	73	—	277
,, ...	,, (West Coast.) Treaties, &c. ..	1819—1890	74	—	278
,, ...	,, (Algeria.) Notes	1830—1844	75	—	287
,, ...	Ratification. "Brussels Act"	2 Jan., 1892	21	—	98–99
,, ...	And Belgium. See Belgium and France.				
,, ...	And Comoro Islands. French Protectorate, &c.	26 June, 1886	77	—	292
,, ...	,, Convention. Anjouan (Johanna) French Protectorate, &c.	21 Apr., 1886	76	—	291
,, ...	,, Ditto. Amendment	15 Oct., 1887	77	—	292
,, ...	And Congo. French Right of Pre-emption..	April—May, 1884	46	—	207
,, ...	,, " "	April, 1887	50	—	215
,, ...	,, Recognition of Association. Boundaries, &c.	5 Feb., 1885	47	—	209
,, ...	,, Private Stations and Properties ..	5 Feb., 1885	48	—	212
,, ...	,, Manyanga Region	22 Nov., 1885	49	—	213
,, ...	,, Oubangi Region ..	29 Apr., 1887	51	—	217
,, ...	,, Right of Pre-emption. Congo Free State	5 Feb., 1895*	App. 30	—	1059
,, ...	Boundaries	14 Aug., 1894	App. 21	—	1021
,, ...	And Danakils. Cession of Obock, &c.	11 Mar., 1862	68	2	269
,, ...	,, Somali Coast and Tajourah	2—9 Feb., 1888	App. 3	—	976
,, ...	And Germany. Limits ..	24 Dec., 1885	78	—	293
,, ...	,, Do. ..	1 Feb., 1897	79	—	297
,, ...	,, Territorial Possessions of Zanzibar. Island of Mafia ..	17 Nov., 1890	App. 7	—	985
,, ...	,, Cameroons. French Congo. Lake Chad. (Map)	4 Feb,, 1894 15 Mar., 1894	App. 14	—	999

* Treaty not yet ratified (1 Feb., 1896).

ALPHABETICAL INDEX.

Name of Country, Place, &c.	Subject.	Date of Treaty or other Document.	No. of Doc.	Art.	Page.
France ..	And Great Britain. Treaties, &c.	1783—1893	103—118	—	539—589 and App. 970
,, ...	Boundary. North and East of Sierra Leone	21 Jan., 1895	App. 29	—	1048
,, ...	And Liberia. Boundaries, &c.	1892—1893	164	—	777
,, ...	And Madagascar. Protection, &c... ..	1868—1885	165—168	—	788—795
,, ...	And Morocco. Boundaries	10 Sept., 1844	170	—	802
,, ...	,, ,, ,,	18 Mar., 1845	171	—	803
,, ...	And Ouatchia. Protection.	10 June, 1885	74	—	284
,, ...	,, French Protectorate	20 Aug., 1885	Note	—	47
,, ...	And Portugal. Boundaries	12 May, 1886	80	—	298
,, ...	,, British Protest against ditto ..	13 Aug., 1887	—	—	325
,, ...	And Zanzibar. Independence of Sultan .. See also Great Britain and France.	17 Nov., 1844	195	—	927
Free Trade Zone	"Berlin Act" (Map) ..	26 Feb., 1885	17	1	24
,,	German Colonization Society	6 Mar., 1885	31	—	304
,,	Application of ditto to British and German Spheres within ..	1 July, 1890	129	8	648
,,	Settlements in	1 July, 1890	129	8	649
,,	Portuguese Reservations. Spheres of Influence (not ratified) ..	20 Aug., 1890	148	—	720
,,	Zanzibar	22 June, 1892	App. 11	—	903
French Congo .	See Congo, French.				
French Guinea	Trade. Open Roads, &c.	21 Jan., 1895	App. 29	—	1054
French Soudan	France and Emir-el Mumenin. Boundary .. See also Fouta Djallon.	23 Mar., 1887	74	—	285
Frere, Mt. ..	Pondoland. See Great Britain (Cape Colony).				
Frio, Cape ..	See Germany and Africa (South-West Coast).				
Fugitive Slaves	Freedom. "Brussels Act"	2 July, 1890	18	64	76
,,	On Ships of War. Freedom. Do.	2 July, 1890	18	28	64
,,	Witu	20 Mar., 1891	34	—	163

ALPHABETICAL INDEX.

Name of Country, Place, &c.	Subject.	Date of Treaty or other Document.	No. of Doc.	Art.	Page.
Gaikas..	(Kaffraria.) See Great Britain (Cape Colony).				
Galabat	See Kalabat.				
Galam..	Fort. Ceded by Great Britain to France..	3 Sept., 1783	103	9	539
Galekaland	See Great Britain (Cape Colony).				
Galla..	See Abyssinia, &c.				
Gallaland	Confines. British and German Spheres of Influence.. ..	1 July, 1890	129	1, § 2	644
Gallinas	(Sierra Leone). Cession to Great Britain with half a mile inland..	30 Mar., 1882	100	—	516
,,	,, Boundaries ..	18 May, 1885	100	—	524
Gama..	See Great Britain (Gold Coast).				
Gambia	Guarantee to Great Britain..	3 Sept., 1783	103	10	539
,,	Cession to Great Britain..	15 June, 1826	90	2	367
,,	Ditto. One mile inland on right bank ..	15 June, 1826	90	3	368
,,	,, ,,	5 Jan., 1832	90	—	372
,,	French Vessels trading to Albreda	15 June, 1826	90	2	369
,,	French Trade	7 Mar., 1857	106	3	545
,,	Union with Sierra Leone..	19 Feb., 1866	90	—	381
,,	Separate Colony	28 Nov., 1888	90	—	383
,,	Territory north of, British and French (Map)	10 Aug., 1889	110	1	558
,,	Territory south of, British and French (Map)	10 Aug., 1889	110	7 Annex 2, § 4	558 565
,,	Notes on the Gambia ..	1783—1894	90	—	365
Gambo-Kona..	Concession. Danakils to Italy	9 Dec., 1888	12	12	12
Gash	British and Italian Spheres	15 April, 1891	13	12	12
Gera	Egyptian Evacuation ..	Jan., 1884	67	—	263
German Colonization Society	Treaties. Native Chiefs..	Oct.—Nov., 1884	81	—	303
,,	(East Coast). Charter ..	17 Feb., 1885	81	—	303
,,	Treaties, Witu and Somali	1885—1887	82	—	312
,,	And Zanzibar ·	1888—1890	198— 200	—	933— 943
,,	Surrender of Concessions to Germany ..	1 July, 1890	129	11	650
German S. W. Africa Company	A Corporate Body ..	13 April, 1885	83	—	318
Germany	Laws. Jurisdiction, &c., in Protectorates ..	1879—1891	—	—	322

ALPHABETICAL INDEX.

Name of Country, Place, &c.	Subject.	Date of Treaty or other Document.	No. of Doc.	Art.	Page.
Germany	Occupation. Territories west of Zanzibar	6 Mar., 1885	82	—	305
,,	Accession. Declaration. Great Britain and France, 1862. Independence. Zanzibar	29 Oct., 1886	123	—	618
,,	Cession by Zanzibar to Germany of coast, Wanga to Rovuma, and Island of Mafia	27—28 Oct., 1890	198	—	940
,,	And Africa. (East Coast) Charter. German Colonization Society	17 Feb., 1885	81	—	303
,,	,, Notes on Protectorates, Treaties, &c.	1884—1890	81, 82	—	303—305
,,	,, (South-West Coast.) Notes on Protectorates. Namaqualand, Damaraland, &c.	1884—1892	83	—	317
,,	,, (West Coast.) Notes on Protectorates	1884—1890	84	—	320
,,	And Congo Free State. Recognition	8 Nov., 1884	52	—	219
,,	And France. Boundaries, &c.	1885—1887	78, 79	—	293—297
,,	,, Madagascar. Territorial possessions of Zanzibar. Island of Mafia	17 Nov., 1890	App. 7	—	985
,,	,, Cameroons. French Congo. Lake Tchad	4 Feb., 1894	App. 13	—	998
,,	And Great Britain. Boundaries, &c.	1885—1894	119—133	—	596—658 and App.
,,	And Portugal. Boundaries	30 Dec., 1886	85	—	323
,,	,, British Protest against ditto	13 Aug., 1887	85	—	325
,,	,, Spheres of Influence. East Africa. Kionga, &c.	Sept., 1894	App. 23	—	1024
,,	And Zanzibar. Consular Jurisdiction, &c.	20 Dec., 1885	197	—	530
,,	See also Kwyhoo, Patta, Manda, Witu, &c.				

ALPHABETICAL INDEX.

Name of Country, Place, &c.	Subject.	Date of Treaty or other Document.	No. of Doc.	Art.	Page.
Gig-giga	British and Italian Spheres	5 May, 1894	136*	1	670
Gildessa	British and French Spheres	2–9 Feb., 1882	App. 3	1	976
,,	British and Italian Spheres	5 May, 1891	136*	1	670
Girrhi Tribes..	British and Italian Spheres	5 May, 1894	136*	1	670
Gobad ..	And France. See Obock.				
Godomey	(Dahomey.) Annexed to France	3 Dec., 1892	65	—	248
Gold Coast	Danish Possessions ceded to Great Britain	17 Aug., 1850	66	—	256
,,	Exchange. Forts and Possessions. Great Britain and Netherlands	5 Mar., 1867	137	—	675
,,	Dutch Forts and Possessions ceded to Great Britain	25 Feb., 1871	137	—	676
,,	Gold Coast separate Colony	13 Jan., 1886	91	—	396
,,	British and French Limits	10 Aug., 1889	110	3 Annex 1	559 567
,,	,, ,, ,,	26 June, 1891	114	4	574
,,	,, ,, ,,	12 July, 1893	118	—	589
,,	Treaties, &c. Great Britain	1831—1891	91	—	385
,, -	British and German Spheres	April—June, 1885	119	—	596
,,	,, ,, ,,	July—Aug., 1886	122	—	612
,,	,, ,, ,,	1 July, 1890	129	4, § 2	647
,,	,, ,, ,,	14 April, 1893	131	—	654
,,	,, in the Interior	15 Nov., 1893	133	—	658
,,	Notes on the Gold Coast..	1831—1891	91	—	388
,,	Gold Coast Colony. East of Volta, and Togoland. Customs Union	24 Feb., 1894	Note	—	661
Gorée ..	Island. Restored to France	3 Sept., 1783	103	9	539
Goubout-Kharab	See Gubbed-Kharab.				
Grain Coast ..	Renunciation by France..	8 Dec., 1892	164	3	785
,,	Liberian Sovereignty. West of Cavally River	8 Dec., 1892	164	3	785
Grand Comoro	See Comoro Islands.				
Grand Sesters	Basin. Liberian Possession..	8 Nov., 1892	164	1, § 2	784
Great Britain	And Abyssinia. See Abyssinia, Ethiopia, Shoa.				

ALPHABETICAL INDEX.

Name of Country, Place, &c.	Subject.	Date of Treaty or other Document.	No. of Doc.	Art.	Page.
Great Britain	And Africa (East Coast), Witu, &c.	19 Nov., 1890	86	—	327
,,	,, British Protectorate North of the Tana..	31 July, 1893	160	—	770
,,	Basutoland	1843—1891	87	—	329
,,	Bechuanaland (British), Bechuanaland, and the Kalahari	1884—1893	88	—	333
,,	And Belgium (Congo)	16 Dec., 1884	53—54	—	221—223
,,	Cape Colony	1652—1887	89	—	339
,,	And Congo Free State. Recognition	16 Dec., 1884	53—54	—	221—223
,,	And Denmark	17 Aug., 1850	66	—	256
,,	And France. Treaty. West Coast of Africa. Senegal, &c.	3 Sept., 1783	103	—	539
,,	,, Treaty. Restoration of certain Colonies to France	30 May, 1814	104	—	540
,,	,, Prussian Award. Portendic Claims	30 Nov., 1843	105	—	541
,,	,, Convention. Portendic and Albreda	7 Mar., 1857	106	—	544
,,	,, Declaration. Independence. Muscat and Zanzibar. (Modified 5th Aug., 1890)	10 Mar., 1862	107	—	547
,,	,, Notes. Tunis. French Protectorate. Treaty Rights of Great Britain and Foreign Powers. Bizerta, &c........	$\frac{16}{20}$ May, 1881	108	—	548
,,	,, Convention. Territorial Limits north of Sierra Leone.....	28 June, 1882	109	—	554
,,	,, Exchange of Notes. Spheres of Influence. Tajourah. Somali Coast ..	$\frac{2}{9}$ Feb., 1888	—	—	970
,,	,, Arrangement. Senegambia to Gold Coast. Slave Coast, Gambia, Sierra Leone, Assinee, Porto Novo..	10 Aug., 1889	110	—	558
,,	,, Exchange of Notes. Approval of Arrangement of 10th August, 1889, ditto	$\frac{2}{19}$ Nov., 1889	111	—	568

ALPHABETICAL INDEX.

Name of Country, Place, &c.	Subject.	Date of Treaty or other Document.	No. of Doc.	Art.	Page.
Great Britain	And France. Declaration. British Protectorate. Zanzibar and Pemba. French Recognition. Modification of Declaration of 10th March, 1862 ..	5 Aug., 1890	112	—	570
,,	,, Declaration. Madagascar. French Protectorate. French Sphere of Influence south of her Mediterranean Possessions. Line from Saye, on the Niger, to Barruwa, on Lake Tchad	5 Aug., 1890	113	—	571
,,	,, Agreement. British and French Spheres of Influence. Niger Districts. Boundary Commission ..	26 June, 1891	114	—	573
,,	,, Exchange of Notes. Ivory Coast. Extension of French Territory to Frontier of Liberia ..	1891—1892	115	—	575
,,	,, Notes. African Boundary Arrangements	1885—1893	116	—	582
,,	,, Joint Report. British and French Commissioners, Panchang and Niambuntang (Gambia)	8 May, 1893	117	—	588
,,	,, Arrangement. Boundary. British and French Possessions on the Gold Coast	12 July, 1893	118	—	589
,,	,, Boundary. North and East of Sierra Leone	21 Jan., 1895	App. 29	—	1048
,,	Gambia	1788—1889	90	—	363
,,	And Germany. Arrangement. Spheres of Action. Gulf of Guinea, Cameroons, Ambas Bay, St. Lucia Bay, &c. ..	29 Apr., 16 June, 1885	119	—	596

ALPHABETICAL INDEX.

Name of Country, Place, &c.	Subject.	Date of Treaty or other Document.	No. of Doc.	Art.	Page.
Great Britain	And Germany. Procès-Verbal. Great Britain, France, and Germany, Maritime, Littoral and Continental Possessions of Sultan of Zanzibar..	9 June, 1886	120	—	605
,,	,, Protocol. British Claims in German Protectorates in South-west Africa, Penguin Islands ..	15 July, 1886	121	—	608
,,	,, Suppl. Arrangement. Spheres of Action. Gulf of Guinea, River Benue, Yola ..	27 July, 2 Aug., 1886	122	—	612
,,	,, Agreement. Sultan of Zanzibar's Sovereignty over Islands and Territories on East African Mainland. Spheres of Influence. Adhesion of Germany to Declaration between Great Britain and France of 10th of March, 1862 ..	29 Oct., 1 Nov., 1886	123	—	615
,,	,, Adhesion. Sultan of Zanzibar to Anglo-German Agreement of October-November, 1886	4 Dec., 1886	124	—	622
,,	,, Agreement. Establishment of Trading Stations within Spheres of Influence in East Africa ..	Mar., 1887	125	—	623
,,	,, Agreement. Discouragement of Annexations in rear of Spheres of Influence in East Africa ...	July, 1887	126	—	625
,,	,, Joint Recommendations. British and German Commissioners. Spheres of Influence. Interior of Togoland. Volta	Dec., 1887	127	—	628

ALPHABETICAL INDEX.

Name of Country, Place, &c.	Subject.	Date of Treaty or other Document.	No. of Doc.	Art.	Page.
Great Britain	And Germany. Award. Belgium. Difference between British East African Company and German Witu Company. Farming of Customs. Administration of Island of Lamu ..	17 Aug., 1889	128	—	630
,,	,, Agreement. Spheres of Influence. East, West, and South-west Africa.. ..	1 July, 1890	129	—	642
,,	,, Protocol. Boundaries. British and German Spheres of Influence. Lake Jipé, Wanga, &c. ..	$\frac{27 \text{ Oct.}}{24 \text{ Dec.}}$, 1892	130	—	652
,,	,, Agreement. Boundary. Gulf of Guinea. Rio del Rey	14 Apr., 1893	131	—	654
,,	,, Agreement. Boundary. From Umba River to Lake Jipé and Kilimanjaro ..	25 July, 1893	132	—	656
,,	,, Agreement. Boundaries. From Gulf of Guinea into the Interior	15 Nov., 1893	133	—	658
,,	Gold Coast	1831—1891	91	—	385
,,	And Italy. Declaration. Abolition of British Consular Jurisdiction at Massowah ..	17 Aug., 1888	134	—	664
,,	,, Protocol. British and Italian Spheres of Influence in Eastern Africa. River Juba to Blue Nile ..	24 Mar., 1891	135	—	665
,,	,, Protocol. British and Italian Spheres of Influence. Ras Kasar to Blue Nile..	15 Apr., 1891	136	—	667
,,	,, Agreement. Spheres of Influence. Eastern Africa. Somali, &c.	5 May, 1894	136*	—	669
,,	Lagos	1851—1891	92	—	403
,,	And Liberia. See Liberia.				
,,	And Madagascar	27 June, 1865	169	—	796
,,	And Morocco. Purchase of Property of North-West Africa Company. Cape Juby	13 Mar., 1895	App. 33	—	1062

ALPHABETICAL INDEX.

Name of Country, Place, &c.	Subject.	Date of Treaty or other Document.	No. of Doc.	Art.	Page.
Great Britain	Natal	1843—1885	93	—	433
,,	And Netherlands. Possessions formerly held by the Netherlands on the West Coast of Africa. Notes	1814—1871	137	—	672
,,	Niger. Notes on the Niger Districts and Niger Coast Protectorate	1882—1893	94	—	440
,,	,, Notification. British Protectorate over Niger Districts	5 June, 1885	95	—	445
,,	,, Royal Charter. "National Africa Company"	10 July, 1886	96	—	446
,,	,, Notification. British Protectorate over Niger Districts. "National Africa Company," now called the "Royal Company"	18 Oct., 1887	97	—	449
,,	,, List. Treaties with African Chiefs	1884—1892	98	—	450
,,	,, Notification. Oil Rivers' Protectorate to be called the "Niger Coast Protectorate"	13 May, 1892	99	—	480
,,	And Portugal. Additional Convention. Portuguese Limits on East and West Coasts of Africa	28 July, 1817	138	—	683
,,	,, Protocol. Dispute respecting Sovereignty over Island of Bulama to be referred to Arbitration	13 Jan., 1869	139	—	684
,,	,, Award. United States. Sovereignty of Portugal over Island of Bulama	21 Apr., 1870	140	—	688
,,	,, Notification. Portuguese Occupation of Island of Bulama	1 Oct., 1870	141	—	691
,,	,, Protocol. Dispute respecting Delagoa Bay (Lourenço Marques) to be referred to Arbitration	25 Sept., 1872	142	—	693
,,	,, British Case. Conflicting Claims to Delagoa Bay (Lourenço Marques)	Sept., 1873	143	—	694

ALPHABETICAL INDEX.

Name of Country, Place, &c.	Subject.	Date of Treaty or other Document.	No. of Doc.	Art.	Page.
Great Britain	And Portugal. British Case. Exchange of Notes. Non-cession of Territory to whichever Power awarded Award. President of French Republic.	June, 1875	144	—	697
,,	,, Portuguese Sovereignty over Delagoa Bay (Lourenço Marques)	24 July, 1875	145	—	701
,,	,, Protocol. Ratification of Convention between Portugal and South African Republic of 11th December, 1875 (with Explanatory Notes)	7 Oct., 1882	146	—	704
,,	,, Treaty. Portuguese Boundary on West Coast of Africa. Navigation of Rivers Congo and Zambesi.	26 Feb., 1884	147	—	713
,,	,, Convention. Spheres of Influence. Navigation of the Zambesi, &c. (not Ratified, but see Agreement of 14th November, 1890) ..	20 Aug., 1890	148	—	715
,,	,, Agreement. Modus Vivendi. Spheres of Influence. Navigation of the Zambesi and Shiré, &c. Note	14 Nov., 1890	149	—	728
,,	,, Treaty. Spheres of Influence. East and Central Africa ..	11 June, 1891	150	—	731
,,	,, Agreement. Modus Vivendi. Spheres of Influence North of the Zambesi ..	Mar.—June, 1893	151	—	743
	See also Portugal and Great Britain.				
,,	Sierra Leone	1788—1892	100	—	481
,,	Tongaland. Treaty. British Protection, Boundaries, &c. ..	6 July, 1887	101	—	529
,,	,, British Sovereignty	23 April, 1895	App. 34	—	1067
,,	,, ,, ,,	11 June, 1895	App. 35	—	1068
,,	And South African Republic. (Transvaal)	—	179	—	837

ALPHABETICAL INDEX.

Name of Country, Place, &c.	Subject.	Date of Treaty or other Document.	No. of Doc.	Art.	Page.
Great Britain	And Zanzibar. Treaty. Exterritoriality. Consular Jurisdiction	30 April, 1886	152	—	751
,,	,, Agreement. Limits of Sultan's Dominions. Islands of Zanzibar, Pemba, Lamu, and Mafia. Kau, Kismayu, Warsheikh, Dar-es-Salaam, Pangani, Kilimanjaro District, Witu, &c. British and German Spheres of Influence	3—4 Dec., 1886	153	—	754
,,	,, Agreement. Lease for 5 years of Sultan's Dominions to British East Africa Company, Kipini to Mruti (except Witu), Lamu, Manda, Patta, Kiwhyu, Kismayu, Brava, Meurka, Magadisho, and Warsheikh..	31 Aug., 1889	154	—	760
,,	,, Provisional Agreement. British Protectorate over Zanzibar Dominions (Succession, &c.)	14 June, 1890	155	—	763
,,	,, Notification. British Protectorate over Zanzibar Dominions (with exceptions)	4 Nov., 1890	156	—	766
,,	,, Declaration. Exercise of Judicial Powers in Zanzibar by other than Consular Officers	2 Feb., 1891	157	—	767
,,	,, Notification. Free Port of Zanzibar	8 Feb., 1892	App.	—	975
,,	,, Notification. Free Trade Zone..	22 June, 1892	App.	—	976
,,	,, Declaration. Consular Jurisdiction ..	16 Dec., 1892	158	—	768
,,	,, Order in C. Natives of British Protectorates. British Jurisdiction over Zanzibar Subjects. Enforcement of Treaties, &c.	17 July, 1893	159	—	769

ALPHABETICAL INDEX.

Name of Country, Place, &c.	Subject.	Date of Treaty or other Document.	No. of Doc.	Art.	Page.
Great Britain	And Zanzibar. Proclamation. Administration of British Protectorate north of the Tana delegated to Sultan of Zanzibar. Witu, &c.	31 July, 1893	160	—	770
,,	Zululand	1836—1888	102	—	531
Great Scarcies	River. See Scarcies Rivers.				
Greboes ..	(G'deboes). Submission to Liberia	23 Sept., 1893	164	—	786
Griqua Country	See Great Britain (Bechuanaland).				
Griqualand East	Annexation to Xesibeland	25 Oct., 1886	89	—	356
Griqualand West	Annexation to Cape Colony	No. 39, 1877	89	—	348
,,	,, ,, ,,	22 Feb., 1878	89	—	348
,,	Amalgamated with ditto..	15 Oct., 1880	—		
,,	Boundary. South African Republic	3 Aug., 1881	179	—	842
,,	,, ,, ,,	27 Feb., 1884	179	—	846—848
,,	Annexation to Cape Colony	29 July, 1887	89	1	361
Gubbed Kharab	Cession by Tajourah to France	18 Oct., 1884	72	—	276
Gubbi	Italian Administration ..	10 Aug., 1887	0	3	9
Guinea.. ..	France and Portugal Limits	12 May, 1886	80	1	298
Gulf of Guinea	See Great Britain (Gold Coast).				
Hadou	Wells. British and French Spheres	2—9 Feb., 1888	App. 3	1	976
Hahamot ..	See Kor Kakamot.				
Halifax Is. ..	See Ichaboe.				
Hanfilah ..	See Amphila.				
Harrar.. ..	Withdrawal. Egyptian Troops	May, 1884	—	67	263
,, ...	Non-annexation or protection by Great Britain or France ..	2—9 Feb., 1888	App. 3	4	977
,, ...	Action of other Powers ..	2—9 Feb., 1888	App. 3	4	977
,, ...	Caravan Route to Zeila ..	2—9 Feb., 1888	App. 3	5	977
Hartley Hill (Bechuanaland)	British Jurisdiction ..	27 June, 1891	App. 9	—	990
Heligoland ..	And Africa. Arrangement. Great Britain and Germany	1 July, 1890	129	—	642

ALPHABETICAL INDEX.

Name of Country, Place, &c.	Subject.	Date of Treaty or other Document.	No. of Doc.	Art.	Page.
Herero..	And Ovomboland. District. German Protectorate .. .. See also Damaraland.	14 Sept., 1892	83	—	319
Hinterland	British and German views	1884	Note	—	661
,,	Annexation in rear of Spheres of Influence	July, 1887	126	—	625
,,	Togoland. Interior Limits	Dec., 1887	127	—	628
,,	Liberia and France. Limits	8 Jan., 1892	164	1	783
,,	Gulf of Guinea .. .. See also Separate Countries.	15 Nov., 1893	133	—	658
Holland's Bird Island	See Ichaboe.				
Hottentot Bay	British Claims	15 July, 1886	121	3	609
Houbbons ..	French Possessions. North of Sierra Leone ..	10 Aug., 1889	110	2	559
Humbe . ..	Germany and Portugal. Boundary	30 Dec., 1886	85	1	323
Ibadan.. ..	(Lagos). Jurisdiction, &c.	23 July, 1888	92	—	430—431
,, ...	,, ,, ,,	15 Aug., 1893	92	—	432
"Ibea" ..	Imperial British East Africa. See British East Africa Company.				
Ibi	River Benué. British and German Spheres ..	5 June, 1885	95	—	445
Ibu	(Lagos). Non-cession of Territory. Boundaries, &c.	31 May, 1888	88	92	428
Ichaboe and Penguin Islands	British Occupation. Ichaboe Island.. ..	21 June, 1861	89	—	345
,,	Included in British Dominions (not confirmed)	12 Aug., 1861	89	—	345
,,	British Occupation. Penguin Islands ..	5 May, 1866	80	—	345
,,	Annexation of Ichaboe and Penguin Islands to Cape Colony ..	16 July, 1866	89	—	345
,,	Letters Patent. Authority to Cape to annex	27 Feb., 1867	89	—	345
,,	Annexed to Cape Colony (in error)	26 June, 1873	80	—	345
,,	Annexed to ditto under Letters Patent ..	6 July, 1874	89	—	345

ALPHABETICAL INDEX.

Name of Country, Place, &c.	Subject.	Date of Treaty or other Document.	No. of Doc.	Art.	Page.
Ichaboe and Penguin Islands	German Annexation of Coast. Cape Frio to Orange River (except Walfisch Bay)	16 Aug., 1874	83	—	317
,,	German Protectorate, ditto	8 Sept., 1874	83	—	318
,,	Conditional German Protectorate	15 July, 1886	121	—	608
,,	See also Great Britain (Cape Colony).				
Idutywa Reserve	Annexation to Cape Colony	12 June, 1876	89	—	343
	See also Great Britain (Cape Colony).				
Ifé	(Lagos). Non-cession of Territory. Boundaries, &c.	22 May, 1888	92	—	426
Igbessa	(Lagos). Non-cession of Territory. Boundaries, &c.	15 May, 1888	92	—	425
Ikankan	Waterway. Rio del Rey Boundary	14 Apr., 1893	131	1	654
Ilaro	(Lagos). Non-cession of Territory. Boundaries, &c.	21 July, 1888	92	—	429
,,	„ British Protectorate	21 July, 1888	92	—	429
Imperial	British East Africa Company ("Ibea"). See British East Africa Company.				
Import Duties	"Berlin Act"	26 Feb., 1885	17	3, 4	26
,,	"Brussels Act"	2 July, 1890	18	—	88
	See also Separate Countries.				
Inland	Frontiers. See Hinterland				
Insuaim	Ferry. Benin River. See Great Britain (Gold Coast).				
Intoxicating	Liquors. See Spirituous Liquors.				
Inyack	Island. See Delagoa Bay.				
Iombo	Island. Cession to Great Britain	5 June, 1821	100	—	483
,,	Ditto	2 Aug., 1824	100	—	489
Isle of France	(Mauritius). Cession to Great Britain	30 May, 1814	104	8	540
Isles de Los	Cession to Great Britain	6 July, 1818	100	—	485
,,	British Possession	28 June, 1882	109	2	555
Italy	And Assab, Aussa, Danakils, and Shoa. See Abyssinia, &c.				
,,	„ British East Africa Company. List of Treaties, &c.	1889—1891	—	—	107
,,	„ Congo. Recognition. Association	19 Dec., 1884	55	—	227
,,	Protection. Danakil Coast	15 Mar., 1883	5	8	5
,,	Sovereignty. Ditto	9 Dec., 1888	12	2	11

ALPHABETICAL INDEX.

Name of Country, Place, &c.	Subject.	Date of Treaty or other Document.	No. of Doc.	Art.	Page.
Italy	Protectorate. Ditto. (Aussa)	6 Dec., 1889	16	—	18
,, ...	Occupation of Massowah..	3 Feb., 1885	7	1	8
,, ...	Protectorate over Zula ..	2 Aug., 1888	11	—	10
,, ...	Relations. Ethiopia ..	10 Feb., 1889	4	—	4
,, ...	Boundaries. Ethiopia ..	2 May, 1889	13	3	13
,, ...	Ditto, ditto	1 Oct., 1889	14	3	16
,, ...	Foreign Relations Ethiopia	2 May, 1889	13	17	14
,, ...	Recognition. King Menelek of Ethiopia ..	1 Oct., 1889	14	1	16
,, ...	Recognition. Ethiopia. Italian Possessions. Red Sea	1 Oct., 1889	14	2	16
,, ...	Jurisdiction. Ethiopia ..	1 Oct., 1889	14	9	17
,, ...	Conduct. Foreign Affairs. Ethiopia	12 Oct., 1889	15	—	17
,, ...	Protectorate over portions of East Coast of Africa. Notification	19 Nov., 1889	163	—	776
,, ...	Notification. Italian Protectorate over Sultanate of Oppia. (Amended 20 May, 1889)	2 Mar., 1889	161	—	772
,, ...	Notification (Amended). Italian Protectorate over Sultanate of Oppia, &c. ..	20 May, 1889	162	—	774
,, ...	Boundary. The Baraka and Red Sea. Seminomadic Tribes on Frontier	25 June, 1895	App. 37	—	1072
,, ...	And Shoa	1883—1887	6	10	6, 9
,, ...	,, Zanzibar. Consular Jurisdiction, &c. ..	1885—1893	201—204	—	924
,, ...	,, Great Britain. Boundary. Spheres of Influence, &c. ..	1888—1894	134—136	—	663
	See also Beilul, Benadir Ports, Gambo-Kona, Gubbi, Ras Dermah, and Torni.				
Itebu	(Lagos). Non-cession of Territory. Boundaries, &c.	28 May, 1888	92	—	426
Ivory Coast ..	French Protectorates ..	1891—1892	164	—	785
,, ..	France and Liberia. Boundary	8 Dec., 1892	164	—	785
,, ..	Renunciations by Liberia East of Cavally River	8 Dec., 1892	164	3	785

ALPHABETICAL INDEX.

Name of Country, Place, &c.	Subject.	Date of Treaty or other Document.	No. of Doc.	Art.	Page.
Jakri	(Lagos). British Protection..	16 July, 1884	92	—	417
„	„ „ Do. and Sovereignty .. See also Great Britain (Lagos).	5 Feb., 1886	92	—	423
James	Fort (Albreda). Guaranteed to Great Britain	3 Sept., 1883	103	10	539
Jarra	British Sovereignty ..	11 Oct., 1887	90	—	383
Jella-Coffee ..	British Occupation ..	22 June, 1874	91	—	391
Jinnak.. ..	Creek. North of the Gambia. British and French Limits..	10 Aug., 1889	110	1, §1 Annex 2 §§ 1, 2	558 564
Jipé	Lake. British and German Spheres of Influence	29 Oct., 1886	123	3	617
„	„ „ „ ..	1 Nov., 1886	123	3	620
„	„ „ „ ..	3—4 Dec., 1886	153	3	756 759
„	„ „ „ ..	1 July, 1890	129	1, §1	642
„	„ „ „ ..	27 Oct.— 24 Dec., 1892	130	—	652
„	„ „ „ ..	25 July, 1893	132	—	656
Johanna ..	French Protectorate ..	21 Apr., 1886	76	—	291
„ ..	„ ..	26 June, 1886	77	—	292
„ ..	„ See also Comoro. Mohilla. ..	15 Oct., 1887	77	—	292
Jolah	British Sovereignty and Protection	15 Sept., 1887	90	—	382
Juba River ..	Territory North of. See Benadir Ports.				
„ ..	To Kismayu. Coast ..	31 July, 1885	82	—	313
„ ..	Navigation by Italy ..	3 Aug., 1889	27	5	140
„ ..	To Confines of Egypt. British and German Spheres	1 July, 1890	129	2	644
„ ..	Kismayu and Witu. Territories between. British Protectorate	19 Nov., 1890	86	—	327
„ ..	To Blue Nile. British and Italian Spheres	24 Mar., 1891	135	—	665
Jurisdiction ..	See Consular Jurisdiction.				
Kabilai ..	German Recognition. French Rights ..	24 Dec., 1885	78	3	296
Kabinda ..	See Cabinda.				
Kaffraria ..	Annexation to Cape Colony	29 Nov., 1847	100	—	500
„ ..	„ „ „ ..	17 Mar., 1865	89	—	343

4 I.

ALPHABETICAL INDEX.

Name of Country, Place, &c.	Subject.	Date of Treaty or other Document.	No. of Doc.	Art.	Page.
Kaffraria ..	See also Great Britain (Cape Colony).				
Kaffu-Bulloms	Cessions to Great Britain	8—10 Mar., 1827	100	—	496
" "	" "	29 Nov., 1847	100	—	500
Kakaye ..	(Gambia). British Military Post opposite..	29 May, 1827	90	—	369
Kalabat ..	Egyptian Administration. Eastern Soudan ..	30 Nov., 1881	67	—	262
"	Egyptian Evacuation ..	Jan., 1884	67	—	263
Kalahari ..	British Sphere of Influence..	1 July, 1890	129	3	646
	See also Great Britain (Bechuanaland).				
Kambia ..	(Sierra Leone). Limits..	26 Dec., 1851	100	—	502
" ..	" " " ..	10 June, 1861	100	—	506
" ..	And Falabah (Sierra Leone). English Road ..	10 Aug., 1889	110	Annex 1	564
	See also Scarcies, Great.				
Kansallah ..	See Jolah and Fogni.				
Kassala ..	Egyptian Evacuation ..	3 June, 1884	3	2	3
"	Right of Italy to occupy temporarily ..	15 Apr., 1891	136	2	668
"	To Metemma. Free Passage. Italians ..	15 Apr., 1891	136	4	669
Katanu ..	Protection to French ..	10 Aug., 1889	110	4, § 3	562
	See also Great Britain (Gold Coast).				
Katima ..	Rapids	11 June, 1891	150	4	733
		May—June, 1893	151	5	744
" ..	See also Zambesi, Upper.				
Kau	Zanzibar Possessions ..	9 June, 1886	120	7	607
"	" ..	29 Oct., 1886	123	1	616
"	" ..	1 Nov., 1886	123	1	619
"	" ..	3, 4 Dec., 1886	153	3	755—758
Ketu	(Lagos). Non-cession of Territory, Boundaries, &c.	29 May, 1888	92	—	427
"	" British Protectorate	29 May, 1888	92	—	427
Khama's Country	See Bechuanaland.				
Kikounya ..	Zanzibar Possessions ..	9 June, 1886	120	4	606
	See also Zanzibar. Limits.				
Kikuyu ..	British Administration ..	1 July, 1895	App. 36	—	1070
Kilambo ..	River. British and German Spheres of Influence	1 July, 1890	129	1, § 2	643

ALPHABETICAL INDEX.

Name of Country, Place, &c.	Subject.	Date of Treaty or other Document.	No. of Doc.	Art.	Page.
Kilambo	To Congo Free State. Ditto	1 July, 1890	129	1, § 3	643
Kilimanjaro Districts	Treaties. German Colonization Society	1884	82	—	309
,,	Charter. Ditto	17 Feb., 1885	81	—	303
,,	Rival claims. Ditto and Zanzibar	1885	82	—	310
,,	,, ,, ,,	Oct.—Nov., 1886	123	4	617—620
,,	British and German. Limits	Oct.—Nov., 1886	123	3	617—620
,,	,, ,, ,,	24 Dec., 1892	130	4	652
,,	,, ,, ,,	25 July, 1893	132	—	656
,,	Withdrawal of Zanzibar claims	3, 4 Dec., 1886	153	4	756—759
Kilwa Is.	Lake Moero. British Possession	12 May, 1894	App. 16	1b	1009
Kilwa-Kiswani	Bay. Zanzibar Possession	9 June, 1886	120	3	606
,,	German Possession	27, 28 Oct., 1890	— —	Note 3	650
Kilwa-Kivingi	Bay. Zanzibar Possession	9 June, 1886	120	3	606
Kionga..	Germany and Portugal	Sept., 1894	App. 23	—	1024
Kipini..	To Juba River. Coast between	31 July, 1885	82	—	313
,, ..	To Kismayu. (Tana River.) Withdrawal of German Protectorate	1 July, 1890	129	2 11	645 651
,, ..	,, British Protectorate	19 Nov., 1890	86	—	327
,, ..	To Kismayu. Administration by British East Africa Company	20 Mar., 1891	34	—	160
,, ..	To Minengani. River. (Tunghi Bay). Zanzibar Possession	9 June, 1886	120	7	607
,, ..	To Wanga. See Wanga to Kipini.				
,, ..	To Manda Bay. Witu Coast line	3, 4 Dec., 1886	153	5	756
,, ..	Up to Witu. Coast between. See Witu.				
,, ..	To Mruti. Lease. Zanzibar to British East Africa Company (except Witu)	31 Aug., 1889	154	1	760
,, ..	,, Concession. Ditto	4 Mar., 1890	30	—	148
,, ..	,, Ditto. Confirmed	5 Mar., 1891	31	—	150

ALPHABETICAL INDEX.

Name of Country, Place, &c.	Subject.	Date of Treaty or other Document	No. of Doc.	Art.	Page.
Kipini ..	See also Minengani River, Tana River, Tunghi, Witu, and Zanzibar Limits.				
Kisijou	Zanzibar Possession See also Zanzibar Limits.	9 June, 1886	120	4	606
Kismayu	Sovereignty of Zanzibar ..	9 June, 1886	120	8	607
,,	,, ,,	Oct.—Nov., 1886	123	1	616
,,	,, ,,	3, 4, Dec., 1886	153	3	755—758
,,	Joint occupation. British East Africa Company and Italy ..	3 Aug., 1889	27	1	138
,,	,, ,, ,,	18 Nov., 1889	28	—	143
,,	,, ,, ,,	8 April, 1890	202	—	949
,,	Concession. Zanzibar to British East Africa Company	4 Mar., 1890	30	1	148
,,	,, In perpetuity ..	5 Mar., 1891	31	—	150
,,	Excluded from British Protectorate of Zanzibar	4 Nov., 1890	156	—	766
,,	To remain to England. British and Italian Spheres .. .. See also Kipini to Kismayu, Wanga to Kipini, and Wanga to Kismayu.	24 Mar., 1891	135	1	665
Kiwhyu	See Kwyhoo.				
Koki ..	Boundary. Uganda ..	18 June, 1894	App. 17	—	1016
Kordofan	Egyptian Province ..	13 Feb., 1841	67	—	250
,,	Excluded from German Sphere	15 Nov., 1893	133	—	960
Kor Kakamot	(Hahamot). British and Italian Spheres ..	15 April, 1891	136	1	668
Kor Lemsen ..	British and Italian Spheres	15 April, 1891	136	1	668
Kosi Bay ..	South African Republic ..	24 July— 2 Aug., 1890	179	16—18 19	874 875
Kotoku..	Berim River. See Great Britain (Gold Coast).				
Kotonou	Seashore ceded to France .	1864	65	—	249
,,	Territory ceded to France	19 May, 1868	65	—	249
,,	French Protectorate acknowledged	3 Oct., 1890	65	—	253
Kowe ..	(Togoland). In German Sphere	Dec., 1887	127	—	628
Krepi ..	See Crepee.				
Krikor ..	(Gold Coast). Boundaries	12 Nov., 1885	91	—	395

ALPHABETICAL INDEX.

Name of Country, Place, &c.	Subject.	Date of Treaty or other Document.	No. of Doc.	Art.	Page.
Krim Country	(Sierra Leone). Cession of portion to Great Britain	5 June, 1883	100	—	522
,,	,, ,, ,,	21 Oct., 1883	100	—	523
Kubango	River. Boundary. Germany and Portugal.	30 Dec., 1886	85	1	323
Kuilu	River. Congo and Portugal Limits	25 May, 1891	59	—	231
	See also Niadi Quillu.				
Kunene	River. Boundary. Germany and Portugal.	30 Dec., 1886	85	1	323
Kwango	France and Congo. Frontiers..	5 Feb., 1885	47	—	209
,,	,,	24 Mar., 1894	App. 15	—	1004
Kwarra River	See Niger River.				
Kwyhoo	Leased by Zanzibar to British East Africa Company	31 Aug., 1889	154	1	760
,,	Kipini to Point opposite. Sovereignty of Witu	1 July, 1890	129	2	645
,,	Point opposite to Kismayu. German Protectorate withdrawn	1 July, 1890	129	2	645
	See also Lamu.				
Lagos	Grant of land to Church Missionary Society.	1 Mar., 1852	92	—	405
,,	Cession to Great Britain of Port and Island..	6 Aug., 1861	92	—	409
,,	British occupation	6 Aug., 1861	92	—	410
,,	Lagos, &c., united to Sierra Leone (revoked)	19 Feb., 1866	92	—	416
,,	Included in Gold Coast Colony	24 July, 1874	92	—	417
,,	A separate Colony	13 Jan., 1886	92	—	422
,,	British Protectorate. Odi to Benin River	5 Feb., 1886	92	—	422
,,	British and French Spheres	10 Aug., 1889	110	4, § 1	561
,,	Treaties, &c., Great Britain	1851—1891	92	—	403
,,	Notes on Lagos	1851—1891	92	—	405
Lamu	(Or Lamou) Island. Zanzibar Possession	9 June, 1886	120	7	607
,,	,, ,, ,,	Oct.—Nov., 1886	123	1	615—618
,,	,, ,, ,,	3—4 Dec., 1886	153	1	754—757
,,	Belgian Award	17 Aug., 1889	128	—	630
,,	Leased by Zanzibar to British East Africa Company	31 Aug., 1889	154	1	760
,,	Concession. Zanzibar to British East Africa Company (for 50 years)	4 Mar., 1890	30	1	148

ALPHABETICAL INDEX.

Name of Country, Place, &c.	Subject.	Date of Treaty or other Document.	No. of Doc.	Art.	Page.
Lamu ..	Leased by Zanzibar to British East Africa Company. Ditto, ditto. Modified. "In perpetuity" See also Wanga to Kipini, and Zanzibar Limits.	5 Mar., 1891	31	—	150
Leckie ..	Port of Lagos See also Great Britain (Lagos).	7 Feb., 1863	92	—	411
Lemain Island	Gambia. Cession to Great Britain	14 April, 1823	90	—	365
Les Sechelles..	Cession to Great Britain..	30 May, 1814	104	8	540
Liberia..	Independent Republic ..	26 July, 1847	104	—	778
,,	And Great Britain. Recognition of Independence ..	March, 1848	164	—	778
,,	Maryland ..	1831—1856	164	—	780
,,	,, annexed to Liberia ..	19 Feb., 1857	164	—	780
,,	Accession to "Brussels Act" ..	25 Aug., 1892	164	—	786
,,	Submission of Greboes ..	23 Sept., 1893	164	—	786
,,	Submission of Cavalla ..	10 Nov., 1893	164	—	786
,,	And Congo. Most-Favoured-Nation Treatment ..	15 Dec., 1891	56	—	229
,,	And France. Boundaries	8 Dec., 1892	164	—	783
,,	,, Hire of Labourers.	8 Dec., 1892	164	4	785
,,	Extension of French Limits ..	1891, 1892	115	—	575
,,	And Great Britain. Boundaries ..	11 Nov., 1885	164	—	781
,,	And United States. Relations ..	1862—1887	164	—	778
,,	Notes on Liberia and Maryland..	1816—1893	164	—	778
Libreville	See Corisco Bay.				
Limpopo	River. Transit. Persons and Goods ..	11 June, 1891	150	12	737
Lindi Bay	Zanzibar Possession ..	9 June, 1886	120	3	600
,,	German Possession ..	27—28 Oct., 1890	Note	—	650
Little Batanga	German Protectorate ..	15 Oct., 1884	84	—	320
Little Free State	Inclusion in South African Republic	24 July—2 Aug., 1890	179	21	877
Little Popo ..	German Protectorate. French recognition.	24 Dec., 1885	78	2	294
,,	,, ,, ,,	1 Feb., 1887	79	—	297
Llaro ..	British Protectorate. Non-cession of Territory.	21 July, 1888	92	—	429
Loangwa River	See Aroangwa.				
Loco Marsamma	Cessions to Great Britain.	8—10 March, 1827	100	—	496
,,	,, ,, ,,	29 Nov., 1847	100	—	499

ALPHABETICAL INDEX.

Name of Country, Place, &c.	Subject.	Date of Treaty or other Document.	No. of Doc.	Art.	Page.
Loema	Or Louisa Loango. Guinea Coast	12 May, 1886	80	3	300
Lo Magundi	Bechuanaland Gold Fields	27 June, 1891	App. 9	—	990
Lome Harbour	(Togo). German Protectorate	15 Oct., 1884	84	—	320
,,	,, ,, ,,	11 Dec., 1875	176	—	823—824
,,	,, ,, ,,	17 May, 1884	176	—	827
Long Island	See Ichaboe Island.				
Lorenço Marquez	Railway. See also Delagoa Bay.	11 Dec., 1875	176	—	824
Loss, Isle de	See Isle of Loss.				
Lower Umfati	Bechuanaland Gold Fields	27 June, 1891	App. 9	—	990
Luango Luce	See Chiloango.				
Luapula	River. Great Britain and Congo	12 May, 1894	App. 16	1B	1009
Luculla River	Congo Limits	1 Aug., 1885	42	—	198
,,	Congo and Portugal	14 Feb., 1885	58	—	232
,,	,, ,, ,,	25 May, 1891	60	2	237
Lukoja	(Niger). British Protectorate	5 June, 1885	95	—	445
,,	Notes on Lukoja	1866—1869	94	—	441
Lukomo	Island. Lake Nyassa. British	May—June, 1893	151	4	744
Lumi	(Or Lume) River. British and German Spheres	3 and 4 Dec., 1886	153	—	754
,,	,, ,, ,,	1 July, 1890	129	—	441
Lunda	Region. Portugal and Congo. Limits	25 May, 1891	59	—	234
,,	,, ,,	24 Mar., 1894	App. 15	—	1004
Lundi River	And parallel of Fort Charter. British Jurisdiction	27 June, 1891	App. 9	—	990
Lundi and Shashi Rivers	British Jurisdiction between	27 June, 1891	App. 9	—	990
Lunte River	British and Portuguese Spheres	11 June, 1891	150	2	731
Ma Bwetie	Boundary	10 June, 1866	100	—	506
Macbatees	See Scarcies (Great).				
Machinga	British Protectorate	21 Sept., 1889	H.T., xviii.	—	130
Macloutsie River	See Shashi and Macloutsie Rivers.				
Madagascar	French Protectorate. Chiefs. West Coast	26 Feb., 1859	168	—	795
,,	Treaty. British Consular Jurisdiction. Suppression of Piracy and the Slave Trade, &c.	27 June, 1865	169	—	796

ALPHABETICAL INDEX.

Name of Country, Place, &c.	Subject.	Date of Treaty or other Document.	No. of Doc.	Art.	Page.
Madagascar	And France. List. Accessions of Madagascar Territory and Islands	1750—1860	163	—	795
,,	,, Treaty. French Consular Jurisdiction, &c.	8 Aug., 1868	165	—	788
,,	,, Representation of Foreign Relations	17 Dec., 1885	166	—	791
,,	,, Defence of Madagascar Territory	17 Dec., 1885	166	11	792
,,	,, Declaration. Maintenance of Treaties with Madagascar. British Rights confirmed	27 Dec., 1885	167	—	794
,,	France and Great Britain. Declaration. British Rights confirmed	5 Aug., 1890	113	—	571
,,	French Protectorate recognised by Great Britain	5 Aug., 1890	113	—	571
,,	Ditto recognised by Germany	17 Nov., 1890	App. 7	—	985
,,	Great Britain and France. Agreement	5 Aug., 1890	113	—	571
,,	Application to, of "Brussels Act". See also Diego Suarez Bay, and names of other places and Islands occupied by France.	2 Jan., 1892	21	—	99
Mafia	Island. Zanzibar Possession	9 June, 1886	120	4	606
,,	Zanzibar Limits	Oct.—Nov., 1886	123	1	615—618
,,	,, ,,	3—4 Dec., 1886	153	1	754—757
,,	German Possession	27, 28 Oct., 1890	—	Note	650
,,	Ditto recognised by France	17 Nov., 1890	App. 7	—	985
,,	Excepted from British Protectorate of Zanzibar	4 Nov., 1890	156	—	766
Magadisho	See Benadir Ports.				
Mahafales	(Madagascar). French Protectorate	10 Aug., 1859	168	—	795
Mahagi	Port. Lake Albert	12 May, 1894	App. 16	2	1009
Mahela	Point and Factory. British and French Boundary north of Sierra Leone	28 June, 1882	109	1	555
,,	,, ,, ,,	10 Aug., 1889	110	§ 1 Annex 2	556

ALPHABETICAL INDEX.

Name of Country, Place, &c.	Subject.	Date of Treaty or other Document.	No. of Doc.	Art.	Page.
Mahin ..	British Protectorate	24 Oct., 1885	91	—	421
,, ..	,, Sovereignty and Protection ..	5 Feb., 1886	92	—	423
Mahin Beach ..	Cession to Great Britain..	24 Oct., 1885	91	—	418
,, ..	British Sovereignty and Protection ..	5 Feb., 1886	92	—	423
Mahinga Kiswere	See Mchinga Kiswere.				
Mahomadou ..	Town. Liberian Possession ..	8 Dec., 1892	164	1, § 3	784
Makalolo ..	British Protectorate	21 Sept., 1889	H.T., xviii	—	130—132
Malimba ..	German Protectorate	15 Oct., 1884	84	—	320
Malindi ..	Zanzibar Possession. See also Zanzibar Limits.	9 June, 1886	120	7	607
Mambrui ..	Zanzibar Possession	9 June, 1886	120	7	607
Mana Bagroo..	(Sherbro.) Cession to Great Britain	9 Nov., 1861	100	—	509
Manda ..	Island. Cession to German Company. Protest against British East Africa Company's Claim ..	2 Sept., 1885	82	—	312 313
,, ..	Leased by Zanzibar to British East Africa Company ..	31 Aug., 1889	154	1	760
,, ..	Refusal of Germany to recognise East Africa Company ..	20 Dec., 1889	82	—	316
,, ..	Withdrawal of German claim to ..	1 July, 1890	129	2	645
,, ..	British Protectorate ..	19 Nov., 1890	86	—	327
Manda Bay ..	To Kipini. See Kipini to Manda Bay.				
,,	Administration of Islands in, by British East Africa Company for Zanzibar ..	31 Aug., 1889	154	—	760
,, ..	British Protectorate ..	19 Nov., 1890	86	—	327
Mandingo ..	Cessions to Great Britain..	30 Dec., 1825	100	—	495
Manica District	British Jurisdiction ..	27 June, 1891	App. 9	—	990
Manica Plateau	British and Portuguese Spheres ..	11 June, 1891	150	2	733
,,	Ditto. Referred to Arbitration ..	7 Jan., 1895	App. 26	—	1037
Mannah ..	(Liberia.) Purchase of Territories by Great Britain ..	11 Nov., 1885	164	3—4	782
Manyanga ..	Congo Limits ..	1 Aug., 1885	42	—	199
,, ..	Region. France and Congo ..	22 Nov., 1885	49	—	213
Mapoota River	Portuguese Sphere of Influence ..	11 June, 1891	150	3	733

ALPHABETICAL INDEX.

Name of Country, Place, &c.	Subject.	Date of Treaty or other Document.	No. of Doc.	Art.	Page.
Mapoota River	British and Portuguese Boundary. Tongaland..	Sept.—Oct. 1895	App. 38	—	1078
Maputaland ..	British Protectorate	11 June, 1895	App. 35	—	1068
Marimba ..	British South Africa Company. Mining rights	24 Nov., 1894	App. 24	—	1025
Maritime Zone	Congo Basin. "Berlin Act"	26 Feb., 1885	17	1, § 2	24
,,	Slave Trade. "Brussels Act"	2 July, 1890	18	20—21	62-63
Maroti.. ..	See Mruti.				
Maryland ..	Notes on ..	1831—1856	164	—	780
,, ..	Annexed to Liberia	19 Feb., 1857	164	—	780
Mashonaland..	British Jurisdiction	27 June, 1891	App. 9	—	990
Massabe ..	Portugal and Congo. Boundary ..	14 Feb., 1885	58	—	232
Massah ..	Cession of portion to Great Britain	5 June, 1883	100	—	522
Massi-Kessi ..	(Portuguese.) British and Portuguese Spheres	11 June, 1891	150	1	733
,, ..	,, ,, ,,	7 Jan., 1895	App. 26	—	1037
Massowah ..	Egyptian Administration	May, 1865	67	—	259
,, ..	,, ,,	27 May, 1866	67	—	260
,, ..	,, ,,	8 June, 1873	67	—	260
,, ..	,, ,,	2 Aug., 1879	67	—	262
,, ..	,, ,,	30 Nov., 1881	67	—	262
,, ..	,, ,,	2 Dec., 1883	67	—	263
,, ..	Italian occupation	3 Feb., 1885	7	—	8
,, ..	Abolition. British Consular Jurisdiction ..	17 Aug., 1888	134	—	664
Matabeleland..	Boundary. South African Republic ..	3 Aug., 1881	179	—	844
,, ..	,, ...	27 Feb., 1884	179	—	851
,, ..	,, ...	30 July, 1884	179	—	861
,, ..	Order in Council. Limits. British Jurisdiction	18 July, 1894	App. 20	—	1020
Matacong ..	Island. Cession to Great Britain ..	30 Dec., 1825	100	—	496
,, ..	,, ,, ,,	18 April, 1826	100	8	496
,, ..	To belong to France	28 June, 1882	100	2	559
,, ..	Ditto. Confirmed	10 Aug., 1889	110	—	555
Mauritius ..	(Island of France.) Ceded to Great Britain ..	30 May, 1814	104	8	540
Mazoe	River. British and Portuguese Spheres	11 June, 1891	150	2	733
Mchinga Kiswere	Bay. Zanzibar possession	9 June, 1886	120	3	606
,,	German Possession	27, 28 Oct., 1890	—	Note	650

ALPHABETICAL INDEX.

Name of Country, Place, &c.	Subject.	Date of Treaty or other Document.	No. of Doc.	Art.	Page.
Mediation	Disagreements. Treaty Powers. "Berlin Act."	26 Feb., 1885	17	12	30
Mediterranean	Possessions of France. Influence south of..	5 Aug., 1890	113	—	572
Melilla..	Moors on Frontier	6 May, 1845	185	2	893
,, ..	Spanish Jurisdiction	24 Aug., 1859	186	—	894
,, ..	Cession of Territory by Morocco	24 Aug., 1859	186	4	894
,, ..	,, ,,	24 Aug., 1859	186	4	895
,, ..	Neutral ground	26 April, 1860	187	3—6	897—898
,, ..	Limits of Fortress..	30 Oct., 1861	188	4	901
,, ..	,, ,,	5 Mar., 1894	Note	—	902
Mellicourie	River. Cessions to Great Britain	2 May, 1877	100	—	517
,,	French Protectorate recognised by Germany	24 Dec., 1885	78	3	295
,,	French control over	28 June, 1882	109	1	555
,,	,, ,,	10 Aug., 1889	110	2	559
				Annex 1	564
				,, 7	565
,,	Boundary. Great Britain and France	28 June, 1882	109	—	554
,,	,, ,,	10 Aug., 1889	110	—	558
Mendi ..	See Sherbro and Mandi.				
Mercury Island	See Ichaboe.				
Metemma	To Kassala. Free passage. Italians	15 April, 1891	136	4	609
Meurka	Zanzibar limits. See also Benadir Ports.	9 June, 1886	120	8	607
Mfumbiro	(British.) British and German Spheres	1 July, 1890	129	1, § 1	642—644
,,	(British.) British and German Spheres. Amended	12 May, 1894	App. 16	—	1008
Mikindani Bay	Zanzibar possession	9 June, 1886	120	3	606
,,	German possession	27, 28 Oct., 1890	—	Note	650
Milmil ..	British and Italian Spheres	5 May, 1894	136*	1	670
Minengani River	(Tunghi Bay.) To Kipini. Sovereignty. Zanzibar	9 June, 1886	120	1	605
,,	,, ,, ,,	Oct.—Nov., 1886	123	1	615
,,	,, ,, ,,	3 and 4 Dec., 1886	153	1	677
Missionaries ..	See Separate Countries.				
Moero ..	Lake. Congo Free State Boundary	1 Aug., 1885	42	—	199

ALPHABETICAL INDEX.

Name of Country, Place, &c.	Subject.	Date of Treaty or other Document.	No. of Doc.	Art.	Page.
Moero	Lake. Congo Free State Boundary. Amended	12 May, 1894	App. 16	1B	1009
Mohilla	French Protectorate ..	26 June, 1886	77	—	292
,,	See also Comoro, Johanna.				
Molembo ..	Portuguese claim	28 July 1817	138	—	683
Molop	River. British Bechuanaland. Limits ..	8 May, 1891	88	—	337
Mombasa ..	(Mombaze.)				
,, ..	Zanzibar possession ..	9 June, 1886	120	6	607
,, ..	Royal Charter. British East Africa Company..	3 Sept., 1888	25	—	119
,, ..	Projected Railway ..	1 July, 1895	App. 36	—	1071
Monopolies ..	Not to be granted. "Berlin Act"	26 Feb., 1885	17	5	26
,, ..	Certain concessions not so deemed. British South Africa Company	29 Oct., 1889	37	20	179
Moondah ..	River. (Corisco Bay.) Spanish claim. ..	—	180	—	883
Morocco ..	And France. Convention. Boundary. Algeria	10 Sep., 1844	170	—	802
,, ..	,, Treaty. Boundaries	18 Mar., 1845	171	—	803
,, ..	And Spain. Treaties ..	1844—1861	184—187	—	891—901
,, ..	See also Spain and Morocco.				
,, ..	And various countries. Convention. Cape Spartel Lighthouse.. See also Ceuta, Melilla, Great Britain and Morocco, and North-West Africa Company.	31 May, 1866	172	—	808
Mt. Ayliff ..	Annexed to Cape Colony.. See also Great Britain (Cape Colony).	23 Aug., 1886	89	—	355
Mozambique ..	And Angola. Portuguese claim to territories between	12 May, 1886	80	4	300
,, ..	,, ,,	30 Dec., 1886	85	3	324
,, ..	,, British Protest against ditto ..	13 Aug., 1887	—	—	325
Mrima ..	Concession. Zanzibar to British East Africa Company	24 May, 1887	24	—	111
,, ...	Royal Charter. Ditto ..	3 Sept., 1888	25	—	119
,, ...	Concession. Ditto ..	9 Oct., 1888	26	—	127

ALPHABETICAL INDEX.

Name of Country, Place, &c.	Subject.	Date of Treaty or other Document.	No. of Doc.	Art.	Page.
Mruti ..	(Or Maroti.) Leased by Zanzibar to British East Africa Company.	31 Aug., 1889	154	1	760
„ ..	Transfer. British East Africa Company to Italy	18 Nov., 1889	28	—	142
„ ..	Concession. Zanzibar to British East Africa Company	4 Mar., 1890	30	—	148
„	Ditto, 1890, confirmed. See also Kipini to Mruti.	5 Mar., 1891	31	—	150
M'sinje	River. Boundary. Germany and Portugal	30 Dec., 1886	85	2	324
„ ..	„ Spheres of Influence. Great Britain and Germany..	1 July, 1890	129	2	643
„	Spheres of Influence. Great Britain and Portugal	11 June, 1891	150	1, § 1	732
Msovero	See Usagara.				
Mukondokwa..	See Usagara.				
Muni River ..	(Corisco Bay.) Spanish Claim	—	180	—	883
Musardou ..	Liberian Boundary. French possession..	8 Dec., 1892	164	1, § 3	784
Muscat ..	Independence. Zanzibar and Muscat..	10 Mar., 1862	107	—	547
Muslakh Is. ..	See Mussa Island.				
Mussa Island.	Sale to British Government	19 Aug., 1840	178	—	832
„ ..	Included in French Protectorate.,	2—9 Feb., 1888	App. 3	2	976
Mutassa	In British Sphere of Influence. Arrangement. Great Britain and Portugal	11 June, 1891	150	—	733
„ ..	„ „ „	7 Jan., 1895	App. 26	—	1037
Naalah	Town. French Possession. Liberian Boundary..	8 Dec., 1892	164	1, § 3	784
Naloes.. ..	(Sierra Leone.) Boundary.	21 Mar., 1851	100	—	502
Namaqualand.	German Protectorate	15 Aug., 1884	83	—	317
„	„ „	1 July, 1890	129	3	645
„	(Great.) German Protectorate	18 Oct., 1884	83	—	318
„	(Red Nation.) Ditto	2 Sept., 1885	83	—	318
„	(Great.) Bastards of Rehoboth. Ditto	15 Sept., 1885	83	—	318
Natal	Notes on Natal	1843—1885	93	—	434
„	District. British Colony.	12 May, 1843	93	—	434
„	Boundary. Zululand	5 Oct., 1843	93	—	434

ALPHABETICAL INDEX.

Name of Country, Place, &c.	Subject.	Date of Treaty or other Document.	No. of Doc.	Art.	Page.
Natal	Annexation to Cape Colony	31 May, 1844	93	—	435
,, ...	Boundaries, &c.	21 Aug., 1844	93	—	435
,, ...	Cession. Amapondas. Boundary	11 Apr., 1850	93	—	435
,, ...	A Separate Colony ..	12 July, 1856	93	—	435
,, ...	Boundaries	5 Feb., 1858	93	—	436
,, ...	,, ,, ...	5 June 1858	93	—	436
,, ...	Nomansland annexed ..	9 Dec., 1863	93	—	436
,, ...	,, ,, ...	7 Sept., 1865	93	—	437
,, ...	Ditto. " Amaquatis. British Protection ..	10 Dec., 1875	93	—	437
,, ...	And Delagoa Bay. Coast between. Non-acquisition by Germany..	April—May, 1885	119	—	597—599
,,	Treaties, &c. Great Britain	1843—1885	93	—	433
,, ...	Boundary. South African Republic	3 Aug., 1881	179	—	842
,, ...	,, ,,	27 Feb., 1884	179	—	848
National Africa Company	Treaties. Native Chiefs..	Jan.—Nov., 1884	98	—	450
,,	Royal Charter	10 July, 1886	96	—	446
,,	Title changed to Royal Niger Company .. See also Great Britain (Niger).	18 Oct., 1887	97	—	449
Netherlands ..	Cession to Great Britain of the Cape of Good Hope, &c.	13 Aug., 1814	137	—	672
,, ..	Exchange of Possessions on Gold Coast ..	5 Mar., 1867	137	—	674
,, ..	Transfer of ditto to Great Britain	25 Feb., 1871	137	—	676
,, ..	Notes on former Dutch Possessions in Africa	1814—1871	137	—	672
,, ..	And Congo. Recognition of Association ..	27 Dec., 1884	57	—	230
Neutrality ..	Territories and Waters. Congo Basin ..	26 Feb., 1885	17	10—12	29—30
,, ..	Works and Establishments	26 Sept., 1885	17	25	39
,, ..	Congo Free State. Limits.	1 Aug., 1885	42	—	198
,, ..	Ditto. Boundaries altered	22 Nov., 1885	49	—	213
,, ..	,, ,,	29 Apr., 1887	51	—	217
,, ..	,, ,,	25 May, 1891	59	—	234
,, ..	,, ,,	12 May, 1894	App. 16	—	1008
New Calabar ..	British. See Great Britain (Niger).				..
New Republic .	Boundary. Zululand ..	22 Oct., 1886	179	—	860
,, .	Union with South African Republic	14 Sept., 1887	179	—	862

ALPHABETICAL INDEX.

Name of Country, Place, &c.	Subject.	Date of Treaty or other Document.	No. of Doc.	Art.	Page.
Ngami .	Lake. British Sphere of Influence	1 July, 1890	120	3	646
Nguru ..	German Colonization Society. Treaty. Native Chiefs	26 Nov., 1884	82	—	305
,,	Charter. Ditto	17 Feb., 1885	81	—	303
Niadi-Quillou	Congo and United States..	22 Apr., 1884	64	—	244
,,	Congo and Great Britain..	16 Dec., 1884	53	—	221
,,	Congo and France. Frontiers..	5 Feb., 1885	47	—	209
,,	Congo Limits . See also Kuilu.	1 Aug., 1885	42	—	199
Niambuntung	Or Niani Buntang (Gambia). Position	8 May, 1893	117	—	588
Nicol Island ..	German Protectorate	15 Oct., 1884	84	—	320
Niger ..	National Africa Company (Royal Niger Company). Charter See also Great Britain (Niger) and Sokoto.	10 July, 1886	96	—	446
Niger Company	Treaties. Native Chiefs..	1884—1892	98	—	450
Niger Districts and Coast	British Protectorate. Districts and Oil Rivers	11 June, 1885	—	—	47
,, ..	British Protectorate. Districts	5 June, 1885	95	—	445
,, ..	,, ,, ,,	18 Oct., 1887	97	—	449
,, ..	British Protectorate. Coast	13 May, 1893	99	—	479
Niger Coast Protectorate	Notes on Niger Districts and Niger Coast Protectorate	1882—1893	94	—	440
,, ,,	Middle and Upper British and French Spheres	10 Aug., 1889	110	3	559
		26 June, 1891	114	—	573
,, ,,	(Saye.) French Influence	5 Aug., 1890	113	2	572
,, ,,	Basin and Affluents	10 Aug., 1889	110	2	559
,, ,,	Ditto. Secured to France	8 Dec., 1892	164	1, § 4	784
,, ,,	Agreement. Great Britain and Germany	15 Nov., 1893	133	6	661
Niger River ..	Free Navigation. Berlin Act ..	26 Feb., 1885	17	5	26—33, 39—43
,, ..	British Engagements	26 Feb., 1885	17	30	41
,, ..	French ditto	26 Feb., 1885	17	31	42
,, ..	Ditto. Of other Powers..	26 Feb., 1885	17	32	42
,, ..	Protectorate. Prohibition. Alcoholic Liquors ..	18 June, 1892	94	—.	443
Nile ..	Upper Watershed of Basin. British and German Spheres of Influence	1 July, 1890	129	1, § 2	644
,, ...	Irrigation works on the Atbara	15 Apr., 1891	136	3	667

ALPHABETICAL INDEX.

Name of Country, Place, &c.	Subject.	Date of Treaty or other Document.	No. of Doc.	Art.	Page.
Nile	Basin. Spheres of Influence. Great Britain and Congo ..	12 May, 1894	App. 16	1A 2	1008–1009
	See also Blue Nile.				
Nokki	Great Britain and Portugal. Boundaries notified	26 Feb., 1884	147	—	714
,, ...	And Lower Congo. Boundaries..	25 May, 1891	60	3	237
Nomansland ..	Annexed to Natal.. ..	9 Dec., 1863	93	—	436
,, ...	,, ,, ...	7 Sept., 1865	93	—	436
,, ...	Between Umzimkulu and Umtamfuna Rivers annexed to Natal ..	9 Dec., 1863	93	—	436
,, ...	(Transkei.) Annexation to Cape Colony ..	12 June, 1876	89	—	345
	See also Great Britain (Cape Colony).				
,, ...	South - West Coast of Africa	14 Sept., 1892	83	—	319
	See also Great Britain (Cape Colony) and Germany (Africa, South-West Coast).				
Non-cession of Territory	Algiers. France	30 May, 1837	75	13	288
,, ...	Obokh. France and Danakils..	11 Mar., 1862	68	9	270
,, ...	Delagoa Bay. Great Britain and Portugal ..	June, 1875	144	—	697
,, ...	Egypt. Somali (proposed)	7 Sept., 1877	67	2—5	261
,, ...	Tajourah. France ..	21 Sept., 1884	71	5	274
,, ...	South of the Zambesi. Great Britain and Portugal	20 Aug., 1890	148	3—4	717—718
,, ...	,, Mutual rights of pre-emption	11 June, 1891	150	7	735
,, ...	See also Treaties with native Chiefs, under different countries.				
North Bulloms	Treaty. Cession. Tombo Island to Great Britain	5 June, 1821	100	—	488
,, ...	Treaty. Cessions. Bance and other islands to Great Britain ..	2 Aug., 1824	100	—	489
North-West Africa Company	Purchase of Property by Morocco	13 Mar., 1895	App. 33	—	1064
Norway ..	See Sweden and Norway.				
Nossi Bé ..	Madagascar. Cession to France	3 Feb., 1841	68	—	795
Nossi Comba ..	Ditto. Ditto	3 Feb., 1841	168	—	795

ALPHABETICAL INDEX.

Name of Country, Place, &c.	Subject.	Date of Treaty or other Document.	No. of Doc.	Art.	Page.
Nossi-Mitsion	Madagascar. Cession to France	1 June, 1841	168	—	795
Notifications..	Occupations, Protectorates, &c. See Occupations, Protectorates.				
Nougona	(Gold Coast.) British and French Boundary	10 Aug., 1889	110	3	559
Ntombo-Makata	Cataract. Congo. Limits	1 Aug., 1885	42	—	199
Nubia ..	Egyptian Province	13 Feb., 1841	67	—	259
Nunez ..	Rio. French Rights	28 June, 1882	109	3	556
,, ..	Ditto. German Recognition..	24 Dec., 1885	7	8	296
Nyasaland Districts	Not included in Chartered territory	2 Apr., 1891	38	—	185
,,	Notification. British Protectorate	14 May, 1891	173	—	811
,,	Prohibition. Alcoholic Liquors	18 June, 1892	86	—	228
,,	Name changed to "British Central Africa Protectorate"	22 Feb., 1893	173	—	811
,,	Notes on Nyasaland	1891—1893	173	—	811
Nyasa Lake ..	And Tanganyika Lake, to Rovuma River	1 July, 1890	129	1, § 1	643
,,	And Congo State. No Transit Dues	1 July, 1890	129	8	649
,,	Boundary. Germany and Portugal	30 Dec., 1886	85	2	324
,,	Spheres of Influence. Great Britain and Germany	1 July, 1890	129	1	643
,,	Ditto. British and Portuguese	11 June, 1891	150	5	734
,,	Ditto	May—June, 1893	151	10	746
,,	And Zambesi. British and Portuguese Spheres	May—June, 1893	151	1	743
Nyasa and Tanganyika Lakes	German Frontier between. Protection against aggression ..	24 Nov., 1894	App. 24	—	1025
Nyasa-Tanganyika	Plateau. British and German Spheres. (Map.)	1 July, 1890	129	1	643
Nyasa ..	Islands on Lake. British	May—June, 1893	151	4	744
Nyasaland	Excluded from British South Africa Co.'s field of operations ..	Feb., 1891	App. 8	—	987
,,	Definition of Territory ..	Feb., 1891	App. 8	—	987

ALPHABETICAL INDEX.

Name of Country, Place, &c.	Subject.	Date of Treaty or other Document.	No. of Doc.	Art.	Page.
Obock	Cession to France (Map)	11 Mar., 1862	68	2	269
,,	Limits. French notification	25 Dec., 1880	69	—	272
,,	France and Gobad	9 April, 1884	70	—	273
,,	Application to, of "Brussels Act"	2 Jan., 1892	21	—	99
Occupations	Italian. Of Massowah	3 Feb., 1885	7	—	8
,,	African Coasts. Notifications to be made. "Berlin Act"	26 Feb., 1885	17	34	43
,,	Ditto. Return laid before Parliament.. See also Protectorates.	1885—1887	—	—	47
Odi	To Benin River. Coast line. British Protectorate and Sovereignty	5 Feb., 1886	92	—	422
Odzi River	And Portuguese Possessions. British Jurisdiction	27 June, 1891	App.9	—	990
Ogaden	Regions. Trade	5 May, 1894	136*	2	670
Ogbo	British Protectorate	24 Dec., 1884	92	—	418
,,	British Sovereignty	5 Feb., 1886	92	—	423
Ohombela	(Niger.) Independent of Obako. British Protection. Non-conclusion of Treaties with Foreign Powers without British sanction. British Jurisdiction, Boundaries, &c.	29 Mar., 1888	—	H.T. viii	187
Oil Rivers	British Protectorate. "Niger Coast"	5 June, 1885	95	—	445
,,	,, ,,	11 June, 1885	—	—	47
,,	,, ,,	13 May, 1893	99	—	479
,,	Lagos to Rio del Rey. Boundary. British and German Spheres of Influence. See Niger Districts, &c.				
,,	Ditto, ditto. Eastern limit	14 April, 1893	131	—	654
,,	Title changed to Niger Coast Protectorate..	13 May, 1893	99	—	479
Okeadan	(Lagos.) British Protectorate	4 July, 1863	92	—	413
,,	Ditto	17 July, 1863	92	—	415
Okrika	British Protectorate	17 May, 1888	—	H.T. viii	190
Old Calabar	River. (Cross River.) See Great Britain (Niger).				
Ondo	(Lagos.) Non-cession of Territory. Boundaries, &c.	20 Feb., 1889	92	—	432
Opobo	See Great Britain (Niger).				
Oppia	Italian Protectorate	2 Mar., 1889	161	—	772

ALPHABETICAL INDEX.

Name of Country, Place, &c.	Subject.	Date of Treaty or other Document.	No. of Doc.	Art.	Page.
Oppia	Italian Protectorate. See also Italy.	20 May, 1889	162	—	774
Orange Free State	Notes on	1818—1876	174	—	814
,,	British Sovereignty ..	3 Feb., 1848	174	—	814
,,	Orange River Territory ..	22 Mar., 1851	174	—	814
,,	Renunciation of British Sovereignty.. ..	30 Jan.. 1854	174	—	814
,,	British recognition of Independence.. ..	23 Feb., 1854	174	—	814
,,	Boundaries. Basutoland..	12 Feb., 1869	174	—	814
,,	Boundaries	13 July, 1876	174	--	818
,,	Adhesion to Brussels Act	24 Nov.. 1894	18	Note	48
Orange River..	To Cape Frio. German Protectorate (excepting Walfisch Bay)	15 Oct., 1884	83	—	318
,, ..	British and German Spheres of Influence	1 July, 1890	129	3	645
Oron	Rio del Rey. Boundary..	14 April, 1893	131	1	654
Ottoman Dominions	See Turkey.				
Ottoman Law..	Slave Trade. See Turkey.				
Ouatchis ..	(Between Grand Popo and Dahomey.) French Protectorate ..	10 June, 1885	74	—	284
,, ..	,, Notification to Powers	20 Aug., 1885	17	—	47
Oubangi ..	Region. Boundary. France and Congo	29 April, 1887	51	—	217
,, ..	,, ,, ,,	14 Aug., 1894	App. 21	—	1021
Ovomboland ..	And Herero. District. See Herero and Ovomboland.				
Oyo	(Lagos.) Boundaries. Non-cession of Territory, &c.	23 July, 1888	92	—	430
Palma	Part of Lagos .. .. See also Great Britain. (Lagos).	7 Feb., 1863	92	—	411
Panchang ..	Or Pantiang. (Gambia.) Position	8 May, 1893	117	—	588
Pangani ..	Zanzibar Possession .. See also Zanzibar Limits.	9 June, 1886	120	5	606
,, ..	Customs. Proposed lease to German East Africa Company ..	Oct.—Nov., 1886	123	2	616
	,, ,, ,,	3, 4 Dec., 1889	153	2	755
,, ..	,, Agreed to by Sultan	4 Dec., 1889	154	—	758

ALPHABETICAL INDEX.

Name of Country, Place, &c.	Subject.	Date of Treaty or other Document.	No. of Doc.	Art.	Page.
Patta	Leased by Zanzibar to British East Africa Company	31 Aug., 1889	154	1	760
,, ...	Withdrawal of German claim to ..	1 July, 1890	129	2	645
,, ...	British Protectorate ..	19 Nov., 1890	86	—	327
,, ...	Zanzibar Administration.. See also Lamu.	31 July, 1893	86	—	327
Peki	See Nepié.				
Pemba Island .	Zanzibar. Possession ..	9 June, 1886	120	1	605
,, ..	,, ,, ,,	Oct.—Nov., 1886	123	1	615, 618
,, ..	,, ,, ,,	3, 4 Dec., 1886	153	1	754
,, ..	Excluded from Concession to British East Africa Company ..	9 Oct., 1888	26	12	134
,, ..	British Protectorate ..	14 June, 1890	155	1	763
,, ..	Import Duties .. .. See also Zanzibar Limits.	22 June, 1892	App. 11	—	993
Penguin Islands	See Ichaboe.				
Peñón de la Pomera	Spain and Morocco, Frontier	6 May, 1845	185	2	893
,, ..	,, ,, ,,	24 Aug., 1859	186	6	895
,, ..	,, ,, ,,	26 April, 1860	187	5	898
Persia	Accession to Berlin and Brussels Acts ..	3 July, 1890	18	—	48
,, ..	Supervision in Territorial Waters, &c. Slaves	2 July, 1890	18	68	78
Philippolis Griquas	See Great Britain (Cape Colony).				
Plantain Island	Cession to Great Britain..	24 Sept., 1825	100	—	491
,, ..	Limits	4—7 July, 1849	100	—	501
Plateau ..	Manica. Slope of. See Manica.				
,, ..	Nyasa-Tanganyika. British and German Spheres. See Nyasa-Tanganyika.				
Plumpudding..	And Roast Beef Island. See Ichaboe.				
Pocrah.. ..	(Pokrah, or Pokéa). (Lagos.) British Protectorate ..	29 June, 1893	92	—	413
,, ...	,, British Possession.	10 Aug., 1889	110	4, § 1	561
,, ...	,, British and French Spheres	10 Aug., 1889	110	4, § 4	563
,, ...	,, Custom House ..	10 Aug., 1889	110	4, § 5	563
Podor	Fort. Cession by Great Britain to France ..	3 Sept., 1783	103	9	539

ALPHABETICAL INDEX.

Name of Country, Place, &c.	Subject.	Date of Treaty or other Document.	No of Doc.	Art.	Page.
Pomona Mine..	South-West Africa. Claims. British Subjects ..	15 June, 1886	121	5	610
Pondoland ..	Cessions to Great Britain.	17 July, 1878	89	—	346
,, ..	British Protectorate of Coast	5 Jan., 1885	89	—	351
,, ..	British Sovereignty ..	20 Mar., 1894	89	—	362
,, ..	Annexation to Cape Colony See also Great Britain (Cape Colony).	12 June, 1894	89	—	362
Pongola ..	River. Free Navigation..	27 July— 2 Aug., 1890	179	12, 13	872
,, ..	,, Limit. Portuguese Influence	11 June, 1891	150	3	733
,, ..	Boundary. Amatongaland..	23 April, 1895	App. 34	—	1067
,, ..	British and Portuguese Boundary. Tongaland..	Sept.—Oct., 1895	App. 38	—	1075
Ponta Vermella	Congo. Limits	1 Aug., 1885	42	—	198
Portendic ..	Cession by Great Britain to France	3 Sept., 1783	103	9	539
,, ..	British Claims on France. Prussian Award ..	30 Mar., 1843	105	—	541
,, ..	Gum Trade	3 Sept., 1783	103	11	539
,, ..	,, ,, ...	7 Mar., 1857	106	1	544
Porto Novo ..	British and French Spheres	10 Aug., 1889	110	4, § 1 Annex 2 § 1	561 567
,, ..	French Protectorate ..	3 Oct., 1890	65	—	253
Porto Seguro..	German Protectorate. French recognition .	24 Dec., 1885	78	2	294
Portugal ..	And British South Africa Company. Boundary	29 Oct., 1889	37	1	175
,, ..	And Congo Boundaries, &c.	14 Feb., 1885	58	—	232
,, ..	,, Tariff of Imports..	9 Feb., 1891	—	Note	238
,, ..	,, ,, ,,	8 April, 1892	—	Note	238
,, ..	,, Spheres of Sovereignty and Influence. Lunda Region	25 May, 1891	59	—	234
,, ..	,, ,, ,,	24 Mar., 1891	App. 15	—	1004
,, ..	,, Boundaries. Lower Congo	25 May, 1891	60	—	236
,, ..	And Dahomey. Protectorate. Sea Coast..	5 Aug., 1885	65	—	253
,, ..	,, ,, ,,	21 Jan., 1886	65	—	253
,, ..	,, Ditto. Withdrawn	22 Dec., 1887	65	—	253
,, ..	And France. Limits. West Africa. Guinea. Fouta-Djallon. Congo. Angola. Mozambique ., ..	12 May, 1886	80	—	298

ALPHABETICAL INDEX.

Name of Country, Place, &c.	Subject.	Date of Treaty or other Document.	No. of Doc.	Art.	Page.
Portugal	And France. French recognition of Portuguese Rights between Angola and Mozambique	12 May, 1886	80	4	300
,,	,, British Protest against above claims of Portugal..	13 Aug., 1887	—	—	325
,,	And Germany. Possessions and Spheres of Influence	30 Dec., 1886	85	—	323
,,	,, ,, ,,	Sept., 1894	App. 23	—	1024
,,	,, Conditional recognition by Germany of Portuguese Rights between Angola and Mozambique	30 Dec., 1886	85	3	324
,,	,, British Protest against above claims of Portugal..	13 Aug., 1887	—	—	325
,,	And Great Britain. Treaties, &c.	1817—1893	138—151	—	681—743
,,	,, Possessions and Claims. East and West Coasts of Africa	28 July, 1817	138	—	683
,,	,, Non-cession of territory without British consent. Delagoa Bay..	June, 1875	144	—	697
,,	See also Bulama and Delagoa Bay.				
,,	,, Boundary. Congo. Zambesi, &c. (not ratified)	26 Feb., 1884	147	—	713
,,	,, Spheres of Influence in Africa (not ratified), but see note, p. 726..	20 Aug., 1890	148	—	715
,,	,, Navigation. Zambesi, Shiré, and Pungwé..	14 Nov., 1890	149	—	728
,,	,, Territorial Limits..	14 Nov., 1890	149	4	729
,,	,, Spheres of Influence. East and Central Africa ..	11 June, 1891	150	—	731
,,	,, Spheres of Influence. North of the Zambesi	May—June, 1893	151	—	743

ALPHABETICAL INDEX.

Name of Country, Place, &c.	Subject.	Date of Treaty or other Document.	No. of Doc.	Art.	Page.
Portugal	And Great Britain. Arbitration. Boundary. Art. 2, Treaty 11 June, 1891. Manica Plateau	7 Jan., 1895	App. 26	—	1037
,,	,, Frontier. British and Portuguese Possessions. Tongaland	Sept.—Oct. 1895	App. 38	—	1075
,,	,, Swaziland	10 Dec., 1894	App. 25	—	1029
,,	,, Boundary. Tongaland	23 April, 1895	App. 34	—	1067
,,	And South African Republic. Treaty. Boundary Dispute. Bay of Lourenço Marques or Delagoa Bay	29 July, 1869	175	—	882
,,	,, Treaty. Boundary, Delagoa Bay Railway, &c.	11 Dec., 1875	176	—	823—824
,,	,, Lorenço Marquez Railway	17 May, 1884	176	—	827
,,	And Zanzibar. Consular Jurisdiction	23 Oct., 1879	206	—	963
,,	Ratification. "Brussels Act"	30 Mar., 1892	23	—	104
Possession	Island. See Ichaboe.				
Pre-emption	Mutual Rights. British and Portuguese. South of the Zambesi	11 June, 1891	150	7	735
,,	,, France. Eventual Cessions (Congo).	$\frac{22}{29}$ April, 1887	50	—	215
,,	See also France and Congo. ,, France and Liberia	8 Dec., 1882	164	5	785
Protectorates.	African Coasts. To be notified. Berlin Act	26 Feb., 1885	17	34	43
,,	(British.) Basutoland	2 Feb., 1884	57	—	332
,,	,, Bechuanaland and Kalahari	30 Sept., 1885	88	—	335
,,	,, British Central Africa	22 Feb., 1893	173	—	811
,,	,, Katanu	15 Mar., 1884	92	—	417
,,	,, Niger Districts and Oil Rivers	5 June, 1885	95	—	445
,,	,, Niger Coast	11 June, 1885	—	—	47
,,	,, Nyassaland	14 May, 1891	173	—	811
,,	,, Pondoland, Coast	5 Jan., 1885	89	—	351
,,	,, Somali Coast	20 July, 1887	178	—	834
,,	,, Uganda	18 June, 1894	App. 17	—	1016

ALPHABETICAL INDEX.

Name of Country, Place, &c.	Subject.	Date of Treaty or other Document.	No. of Doc.	Art.	Page.
Protectorates..	(British.) Witu, &c. Up to Kismayu	19 Nov., 1890	86	—	327
,,	,, North of the Tana	31 July, 1893	86	—	327
,,	,, Zanzibar	14 June, 1890	155	—	763
,,	,, Zanzibar	4 Nov., 1890	156	—	766
,,	,, Uganda	27 Aug., 1894	App. 22	—	1023
,,	,, Tongaland	11 June, 1895	App. 35	—	1068
,,	,, Tati. British Jurisdiction	27 June, 1891	App. 9	—	990
,,	,, Uganda to Coast, and from Juba River to German Spheres	15 June, 1895	App. 36	—	1069
,,	(French.) Alcatras Island	30 Nov., 1887	—	Note	286
,,	,, Comoro Islands	26 June, 1886	77	—	292
,,	,, Dahomey	3 Dec., 1892	65	—	248
,,	,, Ivory Coast	26 Oct., 1891	164	—	783
,,	,, Madagascar. See Madagascar.				
,,	,, Obock. See Obock.				
,,	,, Ouatchis	20 Aug., 1885	—	—	47
,,	,, Tajourah Gulf. See Tajourah.				
,,	,, Tunis	12 May, 1881	190	—	907
,,	(German.) German Colonization Society	17 Feb., 1885	81	—	303
,,	,, Umba to Rovuma	28 Apr., 1888	—	—	923
,,	,, Witu to Kismayu	22 Oct., 1889	82	—	315
,,	,, Withdrawal from Witu	1 July, 1890	129	2	644
,,	,, East Coast of Africa	1834—1890	82	—	305
,,	,, Continental dominions of Zanzibar, and Island of Mafia, recognized by France	17 Nov., 1890	App. 7	—	985
,,	,, South - West Coast of Africa. Namaqualand. Damaraland Coast, Cape Frio to Orange River	1884—1892	83	—	317
,,	,, West Coast of Africa. Togoland. Cameroons. Slave Coast	1884—1890	84	—	320

ALPHABETICAL INDEX.

Name of Country, Place, &c.	Subject.	Date of Treaty or other Document.	No. of Doc.	Art.	Page.
Protectorates..	(German.) Laws. Jurisdiction in Protectorates	1879—1889	—	—	322
,,	(Italian.) Zula ..	2 Aug., 1886	11	—	10
,,	,, Oppia ..	20 May, 1889	162	—	774
,,	,, Aussa..	6 Dec., 1889	16	—	18
,,	,, Ethiopia. Foreign Relations	12 Oct., 1889	15	—	17
,,	,, East African Coast	19 Nov., 1889	163	—	776
,,	(Portuguese.) Sea Coast of Dahomey	21 Jan., 1886	65	—	253
,,	,, Sea Coast of Dahomey, withdrawn	22 Dec., 1887	65	—	253
,,	(Spanish.) N. W. Coast of Africa ..	9 Jan., 1885	182	—	886
,,	Return laid before Parliament	1885—1887	—	—	47
	See also Spheres of Influence and Separate Countries.				
Protest..	British. Against Portuguese territorial claims in Africa ..	13 Aug., 1887	—	—	325
,, ..	Turkish. Against French occupation of Tunis	15—16 May, 1881	190	—	910—913
Pungwé Bay ..	Free passage. British Spheres ..	11 June, 1891	150	14	740
,,	Construction of Railway by Portugal to British Sphere ..	11 June, 1891	150	14	740
,,	Construction of Road by Portugal to British Sphere ..	11 June, 1891	150	14	741
,,	Landing Places ..	11 June, 1891	150	14	741
,,	Telegraph communications ..	11 June, 1891	150	15	741
,,	Transit over Waterways, &c...	14 Nov., 1890	149	2	728
,,	Ditto ..	11 June, 1891	150	12	737
Quahoo	(Gold Coast.) British Protection ..	5 May, 1888	91	—	398
Quiah ..	Cession of portion to Great Britain ..	2 April, 1861	100	—	505
,, ..	British Quiah. Recognition ..	1 Feb., 1862	100	—	510
,, ..	Retrocession of portion of British Quiah ..	20 Jan., 1872	100	—	512
,, ..	British Sovereignty over British Quiah ..	29 Jan., 1872	100	—	512
,, ..	Boundary. British Quiah	20 Jan., 1872	100	—	512

ALPHABETICAL INDEX.

Name of Country, Place, &c.	Subject.	Date of Treaty or other Document.	No. of Doc.	Art.	Page.
Quillu ..	See Niadi-Quillu and Kuilu.				
Quittah	(Gold Coast.) British Occupation..	22 June, 1874	61	—	391
Rahab River ..	British and Italian Spheres of Influence	15 April, 1891	136	1	668
Railways	Swaziland..	10 Dec., 1894	App. 25	—	1029
,,	Mombasa ..	1 July, 1895	App. 36	—	1071
Ras Ali	To Ras Dumeirah. Cession to France	11 Mar., 1862	68	2	269
,,	To Gubbed Kharab. Cession by Tajourah to France	21 Sept., 1884	71	---	274
,, ..	To Gubbed Kharab. French Annexation.	11 Feb., 1885	178	—	833
Ras Dermah ..	To Ras Rakma. Italian Administration	10 Aug., 1887	9	3	9
Ras Dumeirah.	To Amphila. Italian Sovereignty	9 Dec., 1888	12	3	11
,,	To Ras Ali. See Ras Ali.				
Ras Hafoun ..	Somali Coast. Conditional recognition of Egyptian Jurisdiction ..	7 Sept., 1877	67	—	260
Ras Jiburti ..	To Bunder Ziadeh. Somali Coast. British Protectorate	20 July, 1887	178	—	834
Ras Kasar	Italian Spheres of Influence. Red Sea ..	15 April, 1891	136	—	667
,,	To Blue Nile. British and Italian Spheres	15 April, 1891	136	---	667
,,	Frontier. Italy and Egypt	25 June, 1895	App. 37	—	1072
Recruiting	Great Britain and Congo Free State ..	12 May, 1894	App. 16	—	1008
,,	Great Britain and Netherlands. Gulf of Guinea	2 Nov., 1871	137	—	678
,,	France and Liberia. Ivory Coast	8 Dec., 1892	164	4	785
Red Nation	See Germany (Namaqualand).				
Red Sea	British Jurisdiction. See Egypt.				
,,	Egyptian Boundary. See Egypt.				
,,	French Possessions. See France and Africa (East Coast).				

ALPHABETICAL INDEX.

Name of Country, Place, &c.	Subject.	Date of Treaty or other Document.	No. of Doc.	Art.	Page.
Red Sea	Italian Boundary. See Abyssinia, &c., and Great Britain and Italy.				
,,	Italian Possessions recognized by Ethiopia	1 Oct., 1889	14	2	16
,,	Spanish Coaling Station. See Spain and Italy.				
,,	Turkish Claims. See Turkey.				
,,	Boundary. To the Baraka. See Baraka.				
Religious Liberty	Freedom, &c. "Berlin Act" See also separate Countries.	26 Feb., 1885	17	6	27
Rer Ali	Tribes. British and Italian Spheres	5 May, 1894	136*	—	670
Right of Search	Visit and Detention at Sea. "Brussels Act"	2 July, 1890	18	22—23	63
,,	Verification of Ships' Papers. "Brussels Act" See also Slave Trade.	2 July, 1890	18	42, 44	70, 71
Rio Pongas	(Sierra Leone). Boundaries	17 Jan., 1852	100	—	503
Rio del Rey	Great Britain and Germany. Boundary. Gulf of Guinea	April—June, 1885	119	—	596
,,	,, ,, ,,	July—Aug., 1886	122	—	612
,,	Great Britain and Germany. Boundary. Gulf of Guinea	1 July, 1890	129	4, § 2	647
,,	,, ,, ,,	14 April, 1893	131	4, § 2	654
,,	See also Great Britain (Niger Districts and Niger Coast Protectorate).				
Rivers	Lakes, &c. Free Navigation. "Berlin Act" See also Congo, Niger, and others.	26 Feb., 1885	17	2	25
Rivers of the South	Great Britain and France. See Mellicourie. Nunez River.				
Rode Valley	And Territory (Pondoland.) Annexed to Cape Colony See also Great Britain (Cape Colony).	29 July, 1887	89	—	361
Rodrigues	Cession to Great Britain.	30 May, 1814	104	8	540

ALPHABETICAL INDEX.

Name of Country, Place, &c.	Subject.	Date of Treaty or other Document.	No. of Doc.	Art.	Page.
Rovuma	River. British and German Spheres	9 June, 1886	120	—	605
,,	,, ,, ,,	Oct.—Nov., 1886	123	—	615—618
,,	,, ,, ,,	3 and 4 Dec., 1886	153	3	755—758
,,	,, To Lakes Nyasa and Tanganyika	1 July, 1890	129	1, § 2	643
,,	Boundary. Germany and Portugal	30 Dec., 1886	86	2	324
,,	British and Portuguese Spheres	11 June, 1891	150	1, § 1	732
,,	And Wanga Rivers. Zanzibar Dominions	14 June, 1890	155	3	763
,,	,, British and German Spheres	1 July, 1890	129	1—2	642—643
,,	,, Coast Administration by German East Africa Association	28 April, 1688	198	1	933
Roway	Bay of. Egyptian Boundary. See Egypt.				
Ro Woolah	(Sierra Leone). Boundaries	10 June, 1861	100	—	506
Ruo River	British and Portuguese Spheres	11 June, 1891	150	1	743
Russia	And Congo. Recognition. Association	5 Feb., 1885	61	—	239
Saadani	Zanzibar Possessions See also Zanzibar Limits.	9 June, 1886	120	5	606
Sabderate	British and Italian Spheres	15 April, 1891	136	1	667
Sabi (or Save) River	British and Portuguese Spheres of Influence	11 June, 1891	150	2	733
,,	Transit. Persons and Goods. Great Britain and Portugal	11 June, 1891	150	12	737
Sahara	Desert of. France and Morocco	18 Mar., 1845	171	4	805
,, ,,	Great Britain and France. Spheres of Influence	5 Aug., 1890	113	2	572
Sahué	(Gold Coast). British Possessions	26 June, 1891	91	—	*401
St. John's River	Portendic. British Right to Gum Trade	3 Sept., 1783	103	11	539
,,	,, Amended	5 Mar., 1857	106	1	544
,,	Territory (Pondoland). Payment by Great Britain for See also Great Britain (Cape Colony).	9 Dec., 1886	80	—	353

ALPHABETICAL INDEX.

Name of Country, Place, &c.	Subject.	Date of Treaty or other Document.	No. of Doc.	Art.	Page.
St. Louis	Fort. Cession by Great Britain to France	3 Sept., 1783	103	9	539
Sainte Luce	Island and Port. (Madagascar). French Occupation	11 Nov., 1819	168	—	795
St. Lucia Bay	Cession to Great Britain	5 Oct., 1843	93	—	434
,,	British Occupation	18 Dec., 1884	93	—	437
,,	German Protest	7 Feb., 1885	93	—	437
,,	German Recognition of British Flag	April—May, 1885	119	—	597—599
Sainte Marie	(Madagascar). Cession to France	30 July, 1750	168	—	795
,,	,, Re-occupation	15 Oct., 1818	168	—	795
St. Mary Island.	(Gambia). Cession to Great Britain	4 June, 1827	90	—	370
St. Pedro and Cavally Rivers	Liberian Claim to Territory between	1801, 1802	164	—	783
Sakalavas	(Of Féhéréna, Madagascar). French Protectorate	19 Aug., 1850	168	—	795
San Pedro	River. (Gambia). South of. British and French Limits	10 Aug., 1889	110	1, § 2 Annex 2 § 4	554 568
Samanga Is.	Zanzibar Possessions. See also Zanzibar. Limits.	9 June, 1886	120	4	606
Samoo Bullom	Cession to Great Britain	2 May, 1877	100	—	516
Sandeng	(Gambia.) British Possessions	10 Aug., 1889	110	1, § 2 Annex 2, § 5	558 565
Sandwich Harbour	South - West Africa. Claims. British Subjects	15 July, 1886	121	2	603
,,	Coast Fishery	15 July, 1886	121	2	609
Santa Cruz de la Pequeña	Cession by Morocco to Spain	26 April, 1860	187	8	899
Savi	(Dahomey). Annexed to France	3 Dec., 1892	65	—	248
Saye	(Niger). To Barruwa (Lake Chad). French Influence	5 Aug., 1890	113	2	572
Scarcies River	(Small). Boundaries	23 Aug., 1851	100	—	502
,,	(Great). Kambia. Boundaries	26 Dec., 1851	100	—	502
,,	,, Macbatees. Ditto	26 Dec., 1851	100	—	503
,,	,, ,, ,,	11 Dec., 1861	100	—	503
,,	,, Riverain Inhabitants	22 Jan., 1895	App. 29	—	1057
,,	(Great and Small). British Sovereignty over Waters	10 June, 1876	100	—	514

ALPHABETICAL INDEX.

Name of Country, Place, &c.	Subject.	Date of Treaty or other Document.	No. of Doc.	Art.	Page.
Scarcies River	(Great and Small). Ditto over Islands	10 June, 1876	100	—	515
,,	,, Ditto over Territories bordering on Rivers	10 June, 1876	100	100	515
,,	,, Non-cession of Sovereign Rights	10 June, 1876	100	—	516
,,	,, Cessions to Great Britain	2 May, 1877	100	—	517
,,	,, British and French Limits	28 June, 1882	100	—	554
,,	,, ,, ,,	10 Aug., 1889	110	—	558
,,	,, British Control over..	28 June, 1882	109	1	558
,,	,, ,, ,,	18 Nov., 1882	100	—	520
,,	,, ,, ,,	10 Aug., 1889	110	Annex 2 § 7	566
Seal Island	See Ichaboe.				
Secondee	See Great Britain (Gold Coast).				
Self-denying	Declaration. Great Britain and Congo Free State	12 May, 1894	App.	4	992
Senegal	Cession by Great Britain to France	3 Sept., 1783	103	4	539
Senegambia	British and French Possessions	10 Aug., 1889	110	1	558
,,	German Renunciations. See also Coba, Kabilai.	24 Dec., 1885	78	3	296
Senhit	Egyptian Administration. Eastern Soudan	30 Nov., 1881	67	—	262
,, ..	Evacuation. Ditto	Jan., 1884	67	—	263
Sennaar	Egyptian Province	13 Feb., 1841	67	—	259
Seychelles	Ceded to Great Britain	30 May, 1814	104	8	540
Shari	River. Non-extension of German Influence eastward	15 Nov., 1893	133	4	660
Shark Island	South-West Africa. British Claims	15 July, 1886	121	4	609
Shashi	And Lundi Rivers. British Jurisdiction between	27 June, 1891	App. 9	4	991
,, ..	And Macloutsie Rivers. Disputed Territory between. British Jurisdiction	27 June, 1891	App. 9	1A	990
Shavoe River	Boundary. Great Britain and Germany	1 July, 1890	129	4, § 1	647
Shebar	See Bullom and Shebar.				
Sherbro and Mendi	British Right to collect Customs Dues	21 Dec., 1875	100	—	514
Sherbro	(Sierra Leone). Cession to Great Britain	24 Sept., 1825	100	—	491
,, ..	Annexation to Sierra Leone	3 Oct., 1825	100	—	493
,, ..	Cessions to Great Britain	9 Nov., 1861	100	—	508 509

ALPHABETICAL INDEX.

Name of Country, Place, &c.	Subject.	Date of Treaty or other Document.	No. of Doc.	Art.	Page.
Sherbro	British Sovereign Rights confirmed .. See also Bendoo and Chah.	18 Nov., 1882	100	—	520
Sherbro Island	Cession to Great Britain..	24 Sept., 1825	100	—	491
	,, ,, ,,	9 Nov., 1861	100	—	507
Shiré Highlands	Treaties. See British South Africa Company.				
Shiré River	Free Navigation ..	20 Aug., 1890	148	12	722
,,	,, ,, ,,	14 Nov., 1890	149	1	728
,,	British and Portuguese Spheres of Influence	11 June, 1891	150	1, § 2	732
,,	,, ,, ,,	May—June, 1893	151	1 3	743 744
,,	Free Navigation ..	14 Nov., 1890	149	—	730
,,	,, ,, ,,	11 June, 1891	150	12	737
,,	Transit over Waterways..	14 Nov., 1880	149	2	728
,,	,, ,, ,,	11 June, 1891	150	12	737
Shiré Watershed	British and Portuguese Spheres of Influence	11 June, 1891	150	§ 2	732
Shirwa..	Lake. See Chilwa, Lake.				
Shoa	And Great Britain. Treaty. Friendship, &c. .. See also Abyssinia, &c.	16 Nov., 1841	1	—	2
,,	And Italy. Boundaries, &c. ..	21 May, 1883	6	—	6
,,	,, Italian Consular Jurisdiction ..	21 May, 1883	6	12	7
,,	,, Italian Protection of Subjects of Shoa	21 May, 1883	6	13	7
,,	,, Relations ..	20 Oct., 1887	10	—	9
Sierra Leone	Cession to Great Britain..	22 Aug., 1788	100	—	484
,,	Cession of Bance Island to Great Britain	10 July, 1807	100	—	485
,,	British Jurisdiction in countries adjacent to	13 July, 1850	100	—	501
,,	Union with Gambia, Gold Coast, and Lagos ..	19 Feb., 1866	100	—	511
,,	(North Bank). Cessions to Great Britain ..	2 Aug., 1824	100	—	489
,,	,, ,, ,,	29 Nov., 1847	100	—	499
,,	,, ,, Confirmation of ditto	26 Aug., 1852	100	—	503
,,	Union with Gambia ..	17 Dec., 1874	100	—	513
,,	Territory North of British and French Limits..	28 June, 1882	109	—	554
,,	,, ,,	10 Aug., 1889	110	2 Annex 1	558 564
,,	British West Africa Settlements ..	17 June, 1885	100	—	525
,,	,, ,,	11 Oct., 1887	100	—	525
,,	A Separate Colony ..	28 Nov., 1888	100	—	526

ALPHABETICAL INDEX.

Name of Country, Place, &c.	Subject.	Date of Treaty or other Document.	No. of Doc.	Art.	Page.
Sierra Leone	Treaties, &c. Great Britain and Native Chiefs, &c.	1788—1892	100	—	481
,,	,, ,,	1883—1891	100	—	526
,,	Notes on Sierra Leone	1788—1892	100	—	484
,,	British and French Boundary, to the North and East	21 Jan., 1895	App. 29	—	1048
Sinai	Peninsula. Egyptian Administration	8 Apr., 1892	67	—	265
Sinclair	Island. See Ichaboe, &c.				
Slave Coast	German Protectorate. Recognition by France	24 Dec., 1885	78	2	294
,,	,, ,,	1 Feb., 1887	79	—	297
	See also Porto Seguro, Little Popo, Togo.				
,,	British and French Spheres	10 Aug., 1889	110	4	561
,,	British and French Trading Privileges	10 Aug., 1889	110	4, § 2	562
Slaves	Fugitives on board Ships of War. Freedom	2 July, 1890	18	29	64
,,	Detained on board Native Vessels	2 July, 1890	18	29	64
,,	Fugitives. Freedom. "Brussels Act"	2 July, 1890	18	64	76
Slave Trade	By Land and Sea. Berlin Act	26 Feb., 1885	17	—	20
				6	27
				9	29
,,	Suppression. Somali	2—9 Feb., 1888	App. 3	6	976
,,	Means for counteracting. "Brussels Act"	2 July, 1890	18	—	51
,,	Confirmation of Treaties	2 July, 1890	18	24	63
,,	International Bureau at Zanzibar	2 July, 1890	18	27	64
				62	79
,,	Use of Flag by Native Vessels	2 July, 1890	18	30—61	65—76
,,	Conventions concluded before "Brussels Act" repealed	2 July, 1890	18	96	86
,,	Suppression in Maritime Zone	2 July, 1890	18	20—21	62—63
,,	Right of Search, Visit, &c. See also Right of Search.	2 July, 1890	18	23	63
,,	Ottoman Law	4—16 Dec., 1889	192	—	919

ALPHABETICAL INDEX.

Name of Country, Place, &c.	Subject.	Date of Treaty or other Document.	No. of Doc.	Art.	Page.
Slave Trade ..	See also British East Africa Company. Congo ; — Treaties with Foreign Powers.				
Slavery	Suppression. "Berlin Act"	26 Feb., 1885	17	6	27
,,	Slaves sent to Countries in which recognised "Brussels Act" ..	2 July, 1890	18	62—73	76—79
Small Scarcies	See Scarcies.				
Socotra ..	British Protection ..	23 Apr., 1886	177	—	828
Sokoto ..	National Africa Company. Transfer of Rights to, &c.	1 June, 1885	App. 1	—	972
,, ..	Royal Niger Company. Jurisdiction over Foreigners, &c. ..	15 Apr., 1890	App. 6	—	984
,, ..	Influence of Niger Company	5 Aug., 1890	113	2	572
Somali Coast ..	Notes on the Somali Coast	1840—1894	178	—	832
,, ..	French Possessions. (Danakil Coast) ..	1862—1884	68—73	—	269—277
,,	Abandonment by Egypt ..	May, 1884	67	—	263
,,	French Annexations notified	11 Feb., 1885	71	Note	274
,,	British Protectorate over Native Tribes ..	1884—1886	178	—	834
,,	British Protectorate notified	20 July, 1887	—	Note	47
,,	Ras Jiburti to Bunder Zindeh	20 July, 1887	178	—	834
,,	Egyptian Administration. Conditional recognition	7 Sept., 1887	67 178	1 —	260 833
,,	Non-cession by Egypt to any Foreign Power	7 Sept., 1887	67	2	261
,,	Ditto. Assurance required from Sultan ..	7 Sept., 1887	67	5	261
,,	British and French Limits	2—9 Feb., 1888	App. 3	—	976
,,	British Jurisdiction ..	13 Dec., 1889	67 178	— —	265 835
,,	French Recognition. British Protectorate ..	2—9 Feb., 1888	App. 3	2	976—978
,,	Reservation. Rights of Turkey	2—9 Feb., 1888	App. 3	—	976
,,	British and Italian Spheres of Influence ..	5 May. 1894	136*	—	669
Songwe ..	British and German Spheres	1 July, 1890	129	1, § 2	643
Soombia Soosoos	And Turo. Cessions to Great Britain ..	18 Apr., 1826	100	—	495

ALPHABETICAL INDEX.

Name of Country, Place, &c.	Subject.	Date of Treaty or other Document.	No. of Doc.	Art.	Page.
Soudan	Abandonment by Egypt.. See also Eastern Soudan	Jan., 1884	67	—	263
Soulimaniah	British Possession. North of Sierra Leone	10 Aug., 1889	110	2	559
Soumbuya	See Soombia Soosoos.				
South Africa	Company. See British South Africa Company.				
South-West Africa	See Africa (South-West).				
South African Republic	Independence of Transvaal Boers	17 Jan., 1852	179	—	839
,,	Boundaries, &c. Portugal	29 July, 1869	175—179	—	840
,,	Boundary. Barolongs and Batlapins	17 Oct., 1871	179	—	840
,,	Boundaries. Portugal	11 Dec., 1875	179	—	840
,,	British Ratification ditto..	7 Oct., 1882	146	—	704
,,	Declared to be British Territory	12 Apr., 1877	179	—	840
,,	Part of the British Dominions	29 Sept., 1879	179	—	841
,,	Self-government guaranteed under British Suzerainty	3 Aug., 1881	179	—	841
,,	Boundaries. Griqualand West, Natal, Zululand, Swaziland, Portugal, Matabeleland, and Bechuanaland..	3 Aug., 1881	179	—	841
,,	Ditto, ditto (amended)	27 Feb., 1884	179	—	847
,,	Treaties with Foreign States other than the Orange Free State	27 Feb., 1884	179	—	854
,,	Treaties with Netherlands and Portugal	Feb.—Mar., 1884	179	—	857
,,	Award. South-West Boundary..	5 Aug., 1885	179	—	858
,,	Boundary. New Republic and Zululand	22 Oct., 1886	179	—	860
,,	Treaty with Lo Bengula. Matabeleland	30 July, 1887	179	—	861
,,	Union with New Republic	14 Sept., 1887	170	—	862
,,	Treaty with Great Britain. Boundaries..	11—20 June, 1888	179	—	862
,,	Ditto. Swaziland Affairs	24 July—2 Aug., 1890	179	8	870
,,	Non-extension to North or North-West	24 July—2 Aug., 1890	179	10	871

ALPHABETICAL INDEX.

Name of Country, Place, &c.	Subject.	Date of Treaty or other Document.	No. of Doc.	Art.	Page.
South African Republic	Sovereignty recognized by Great Britain	24 July— 2 Aug., 1890	179	12	872
,,	Notes on the South African Republic (Transvaal)	1852—1890	179	—	839
,,	Frontiers. British and Portuguese Spheres of Influence	11 June, 1891	150	—	731
,,	Boundary. Tongaland	23 Apr., 1895	App. 34	—	1067
,,	And Great Britain. Swaziland	10 Dec., 1894	App. 25	—	1029
Spain	Notes on Spanish Possessions. See also Preface, Vol. i, p. xiii.	1479—1893	180	—	882
,,	And Africa (West Coast). Fernando Po and Annabon. Cession by Portugal	1 Mar., 1778	181	—	882
,,	,, ,,	24 Oct., 1778	181	—	884
,,	,, Protection. (North-West Coast) Cape Blanco to Cape Bojador	9 Jan., 1885	182	—	886
,,	And Congo. Recognition of Association	7 Jan., 1885	62	—	240
,,	And Italy. Spanish Naval Station. Assab Bay (Danakil Coast)	Dec., 1887	183	—	888
,,	And Morocco. Limits. Ceuta	25 Aug., 1844	184	—	891
,,	,, ,, ,,	7 Oct., 1844	184	—	891
,,	,, Limits. Ceuta Larache, Melilla, &c.	6 May, 1845	185	—	893
,,	,, Limits. Jurisdiction. Melilla, &c.	24 Aug., 1859	186	—	894
,,	,, Extension. Jurisdiction. Ceuta	26 Apr., 1860	187	—	897
,,	,, Melilla Boundary	30 Oct., 1861	188	—	901
,,	,, Indemnity, &c. Events near Melilla, Oct.—Nov., 1893	5 Mar., 1894	—	Note	902
,,	Ditto. Indemnity	24 Feb., 1895	App. 32	—	1062
Spartel	Lighthouse. See Cape Spartel.				
Spheres of Influence	"Berlin Act"	26 Feb., 1885	17	—	19
,,	British East Africa Company	3 Sept., 1887	25	1	21
,,	British East Coast of Africa. Notes	1887—1891	36	—	170

ALPHABETICAL INDEX.

Name of Country. Place, &c.	Subject.	Date of Treaty or other Document.	No. of Doc.	Art.	Page.
Spheres of Influence	British South Africa Company. Charter	29 Oct., 1889	37	1	175
,,	British South African Company. Non-extension into, of South African Republic ..	24 July—2 Aug., 1890	38	—	183
,,	Congo. Limits ..	1 Aug., 1885	42	—	198
,,	,, Treaties. Foreign Powers	1884—1894	40—64	—	191
,,	British Protest against ditto	13 Aug., 1887	85	—	325
,,	Great Britain and France. North of Sierra Leone ..	28 June, 1882	109	3	560
,,	,, Gold Coast ..	10 Aug., 1889	110	1	560
,,	,, Slave Coast ..	10 Aug., 1889	110	4	561
,,	,, Somali Coast and Bay of Tajourah	2-9 Feb., 1888	App. 3	—	976
,,	,, South of Mediterranean Possessions..	5 Aug., 1890	113	2	572
,,	,, Middle and Upper Niger and Gold Coast ..	26 June, 1891	114	—	573
,,	Great Britain and Germany. Gulf of Guinea ..	Apr.—June, 1885	119	—	596
,,	,, Interior of Gulf of Guinea ..	July—Aug., 1886	122	—	612
,,	,, East Coast of Africa. Zanzibar..	Oct.—Nov., 1886	123	—	616, 617
,,	,, Trading Stations ..	Mar., 1887	125	—	619, 620
,,	,, Non-annexation in rear of Spheres of Influence ..	July, 1887	126	—	625
,,	,, Togoland. Volta..	Dec., 1887	127	—	628
,,	,, East, West, and South-West Africa..	1 July, 1890	129	1, 3, 7	642, 645, 648
,,	,, East Africa, Lake Jipé, Kilimanjaro, Wanga, &c...	27 Oct.—24 Dec., 1892	130	—	652
,,	,, River Benué, Lake Chad ..	14 Nov., 1893	133	—	660
,,	Great Britain and Italy. River Juba to Blue Nile ..	24 Mar., 1891	135	—	665
,,	,, Blue Nile to Ras-Kasar ..	15 Apr., 1891	136	—	667

ALPHABETICAL INDEX.

Name of Country, Place, &c.	Subject.	Date of Treaty or other Document.	No. of Doc.	Art.	Page.
Spheres of Influence	Great Britain and Italy. Eastern Africa. Somali	5 May, 1894	136*	—	669
,, ,,	Great Britain and Portugal.	20 Aug., 1890	148	2, 4, 8	715
					718
					719
,, ,,	,, North and South of the Zambesi	14 Nov., 1890	149	—	728
,, ,,	,, East and Central Africa	11 June, 1891	150	—	731
,, ,,	,, North of the Zambesi	May—June 1893	151	—	743
,, ,,	Portuguese spheres claimed between Angola and Mozambique	12 May, 1886	80	4	300
,, ,,	Ditto, ditto	30 Dec., 1886	83	3	324
,, ,,	See also Hinterland, and Boundary Arrangements between different Countries.				
Spirituous Liquors	Restriction in Traffic. Gold Coast	10 Aug., 1889	110	3, § 3	561
,, ,,	,, "Brussels Act"	2 July, 1890	18	90—95	84
,, ,,	,, Witu, Nyasaland, Niger Protectorate, &c.	18 June, 1892	86	—	328
,, ,,	,, Benadir Ports	12 Aug., 1892	203	4	953
,, ,,	,, Zanzibar and British East Africa Company	9 Oct., 1888	26	4	130
Stanley Pool	Limits. Belgium and France	5 Feb., 1895*	App. 30	—	1059
Stellaland	Boundary. South African Republic	27 Feb., 1884	179	—	852
Stevenson's Road	British and German Spheres. (Map.)	1 July, 1890	129	1, § 2	643
Suakin	Egyptian Administration	May, 1865	67	—	259
,,	Retention by British Government	6 Feb., 1884	67	—	263
,,	Egyptian Administration	27 May, 1866	67	—	260
,,	,, ,,	8 June, 1873	67	—	260
,,	,, ,,	30 Nov., 1881	67	—	262
,,	,, ,,	22 Mar., 1892	67	—	265
Sulymah River	Gallinas (Sierra Leone)	30 Mar., 1882	100	—	517
Swaziland	Independence recognised	3 Aug., 1881	179	24	847
,,	Ditto	27 Feb., 1884	179	12	856
,,	Ditto	24 July— 2 Aug., 1890	179	1	868
,,	Boundary	3 Aug., 1881	179	—	843
,,	Ditto	27 Feb., 1884	179	—	840

* Treaty not yet ratified (1 Feb., 1896).

ALPHABETICAL INDEX.

Name of Country, Place, &c.	Subject.	Date of Treaty or other Document.	No. of Doc.	Art.	Page.
Swaziland	Award. Orange Free State. South-West Boundary	5 Aug., 1885	179	—	858
,,	Abrogation of Treaty of 27 Feb., 1884 (South African Republic, excepting Two Articles)	8 Nov., 1893	189 Note	—	903 1036
,,	Notes on	1881—1893	189	—	903
,,	Railway across Portuguese Territory to Transvaal	—	146	—	708 710 712
,,	Railway. South African Republic	24 July— 2 Aug., 1888	179	8	870
,,	British and Portuguese Spheres of Influence	11 July, 1891	150	2	732
,,	Convention. Great Britain and South African Republic	10 Dec., 1894	App. 25	—	1029
,,	Boundary. Tongaland	23 Apr., 1895	App. 34	—	1067
Sweden	And Norway and Congo. Recognition of Association	10 Feb., 1885	63	—	242
Tajourah	Treaty. East India Company	14 Dec., 1840	71	—	275
,,	Customs Dues. Trade	7 Sept., 1877	67	2	261
,,	Cession to France. (Map)	21 Sept., 1884	71	—	274
,,	Further Cession to France	1 Oct., 1884	72	—	276
,,	,, ,,	14 Dec., 1884	73	—	277
,,	Gulf. British Recognition. French Protectorate	2-9 Feb., 1888	App. 3	2	976—978
,,	Reservation. Rights of Turkey	2-9 Feb., 1888	App. 3	—	979
Taka	Egyptian Administration. Eastern Soudan	May, 1865	67	—	260
,,	,, ,,	30 Nov., 1881	67	—	262
Takaungo	Zanzibar Possession. See also Zanzibar Limits.	9 June, 1886	120	6	607
Talla	British Possession. North of Sierra Leone	10 Aug., 1880	110	2	559
Tambakka	And Bennah. Boundary. See Bennah and Tambakka.				,,
Tamboukie	See Great-Britain (Cape Colony).				

ALPHABETICAL INDEX.

Name of Country, Place, &c.	Subject.	Date of Treaty or other Document	No. of Doc.	Art.	Page.
Tamisso	North of Sierra Leone. French Possession	10 Aug., 1889	110	2	559
Tana	River to Kipini. See Kipini to Tana River.				
,,	,, To Rovuma River British and German Spheres	3–4 Dec., 1886	153	3	755
,,	,, To territory North of. Withdrawal German Protectorate	1 July, 1890	129	2	645
,,	,, British East Africa. Administration	19 Nov., 1890	86	—	327
,,	,, Withdrawal of British East Africa Company. Territory north of	31 July, 1893	160	—	770
,,	,, To Juba River. British Protectorate delegated to Sultan of Zanzibar	31 July, 1893	160	—	770
,,	,, Administration of British Protectorate	30 Aug., 1893	160	—	770
Tanganyika Lake	Congo Limits. Neutrality	1 Aug., 1885	42	—	199
,,	German Trading Station. River Kilifi	Mar., 1887	125	—	623
,,	And Nyasa Lake. No Transit Dues between	1 July, 1890	129	8	649
	See also Nyasa.				
,,	,, North. Boundary of British and German Spheres between	1 July, 1890	129	8	650
,,	No Transit Dues on Lake	1 July, 1890	129	8	650
,,	To Rovuma River	1 July, 1890	129	1, § 2	643
,,	Spheres of Influence. Great Britain and Congo State	12 May, 1894	App. 16	—	1008
,,	Art. 3 ditto withdrawn	22 June, 1894	App. 18	—	1017
Tanoe	River. See Tendo River.				
Tasso Island	Cession to Great Britain.	24 Sept., 1825	100	—	491
Tati	District. Exclusion from Charter of British South Africa Company	29 Oct., 1889	37 {	2 4 20	176 177 179
,,	Proclamation	27 June, 1891	App. 9	—	990
Taveta	Treaties. German Colonization Society	May–July, 1885	82	—	309
,,	British and German Spheres	Oct.–Nov., 1886	123	3	617–6

ALPHABETICAL INDEX.

Name of Country, Place, &c.	Subject.	Date of Treaty or other Document.	No. of Doc.	Art.	Page.
Taveta..	British and German Spheres	3—4 Dec., 1886	153	4	756 759
,,	,, ,,	1 July, 1890	129	1	642
Tchad ..	Lake. See Chad.				
Telegraphs	Coast and British Sphere. South of the Zambesi..	11 June, 1891	150	15	741
,,	Great Britain and Congo State. North of the Zambesi See also Separate Countries.	12 May, 1894	App. 16	5	1011
Tembé ..	A headstream of the Niger (Niger). British and French Boundary ..	26 June, 1891	114	—	573
Tembé-Counda		26 June, 1891	114	—	574
,,	Boundary. France and Liberia	8 Dec., 1892	164	1, § 3	784
,,	Anglo-French Frontier. Sierra Leone	21 Jan., 1895	App. 11	—	993
Tembuland	(Transkei). See Great Britain (Cape Colony).				
Tendo ..	River (Gold Coast). Custom Houses	10 Aug., 1889	110	3, § 2	560
,, ..	,, Free Navigation ..	10 Aug., 1889	110	3, § 2	560
,, ..	River. Fishing Rights ..	12 July, 1893	118	5	591
,, ..	Lagoon. British and French Boundary ..	10 Aug., 1889	110	3, § 2	560
,,	,, ,, Assinee ..	10 Aug., 1889	Annex	7, § 1	567
Terfaya (in Tekna)	See Cape Juby.				
Territorial Waters	Neutrality. Conventional basin of the Congo..	26 Feb., 1885	17	10	29
,,	Freedom of Navigation during War. Congo, &c.	26 Feb., 1885	17	25	38
,,	Ditto. Niger, &c.	26 Feb., 1885	17	33	42
,,	Zanzibar, Rights of British East Africa Company	24 May, 1887	24	9	116
,,	Fishing Vessels ..	2 July, 1890	18	30	18
,,	Jurisdiction	2 July, 1890	18	42	70
Territory	Non-cession. See Non-cession of Territory.				
Tetuan..	Spanish Evacuation	30 Oct., 1861	188	1	901
Tintingue	(Madagascar). French Occupation..	4 Nov., 1818	168	—	795
Togoland	German Protectorate	5 July, 1884	84	—	320
,,	,, ,, ,,	—	129	Note	646
,,	,, French recognition	24 Dec., 1885	78	2	294
,,	British and German Limits	Dec., 1887	127	—	628
,,	Boundary. Gold Coast (Map.)	1 July, 1890	129	4	646

ALPHABETICAL INDEX.

Name of Country, Place, &c.	Subject.	Date of Treaty or other Document.	No. of Doc.	Art.	Page.
Togoland	Gold Coast, and East of the Volta. Customs Union	24 Feb., 1894	Note	—	661
Tombo ..	See Iombo.				
Tongaland	Non-cession of Territory. Boundaries ..	6 July, 1887	101	—	529
,,	South African Republic. Land for Railway ..	24 July— 2 Aug., 1890	179	—	871
,,	,, Abrogation of Treaty of 1890, except Articles 10 and 24 ..	8 Nov., 1893	189	—	903
,,	,, Treaties to be approved ..	24 July— 2 Aug., 1890	179	—	873
,,	South African Republic.				
,,	,, Kosi Bay, &c. ..	24 July—	179	—	874
,,	,, Customs Union ..	2 Aug.,	179	—	875
,,	,, Treaties ..	1890	179	—	876
,,	Annexation and Incorporation with Zululand ..	23 April, 1895	App. 34	—	1067
,,	Boundary. Swaziland and South African Republic ..	23 April, 1895	App. 34	—	1067
,,	British Protectorate over part ..	11 June, 1895	App. 35	—	1068
,,	British and Portuguese Frontier ..	Sept.—Oct. 1895	App. 38	—	1075
,,	Non-recognition of Concessions to Amatongaland Exploration Company ..	4 Nov., 1895	App. 39	—	1078
Torni ..	Aussa Sea Coast to. Italian Administration ..	10 Aug., 1887	9	3	9
Towe ..	Tribe. (West Togo District.) German Protectorate ..	Dec., 1887	127	—	628
Transkei	See Great Britain (Cape Colony).				
Transvaal	See South African Republic.				
T'Slambie	(Kaffraria). See Great Britain. (Cape Colony).				
Tuli ..	District. British Jurisdiction ..	27 June, 1891	App. 9	—	990
Tunghi Bay ..	To Kipini. Zanzibar Sovereignty ..	9 June, 1886	120	2	606

ALPHABETICAL INDEX.

Name of Country, Place, &c.	Subject.	Date of Treaty or other Document.	No. of Doc.	Art.	Page.
Tunghi Bay	To Kipini. Zanzibar Sovereignty	Oct.—Nov., 1886	123	1	615
,,	,, ,, ,,	3 and 4 Dec., 1886	153	1	674—677
Tunis	Boundaries	23 Feb., 1871	190	—	906
,,	Notes on Tunis	1863—1881	190	—	906
,,	Temporary occupation by France	12 May, 1881	190	2	907
,,	Treaties with Foreign Powers guaranteed by France	12 May, 1881	190	4	908
,,	,, ,, ,,	May, 1881	108	—	548
,,	French Protectorate	12 May, 1881	190	—	907
,,	,, Protest of the Porte	16 May, 1881	190	—	911—913
,,	British Consular Jurisdiction abolished	31 Dec., 1883	190	—	915
Tura	See Soombia Soosoos.				
Turkey	Protest against Treaty between France and Tunis of 1881	16 May, 1881	190	—	911—913
,,	Notes on the Ottoman Dominions	—	191	—	918
,,	Law against Negro Slave Trade	16 Dec., 1889	192	—	919
,,	Claims. Basin of Upper Nile	12 May, 1894	App. 16	—	1008
,,	Reservation. Rights of Turkey. Gulf of Tajoura. Somali Coast	2-9 Feb., 1888	App. 3	—	976
,,	Ditto. Basin of Upper Nile	12 May, 1894	App. 16	—	1011-1012
Turtle Island	(Sierra Leone). Cession to Great Britain	9 Nov., 1861	100	—	507
Uganda	British and German Spheres	1 July, 1890	129	1, § 2, 3	644
,,	British and Italian Spheres	15 April, 1891	136	—	667
,,	British East Africa Company. Treaty. King Mwanga (not ratified)	3 Mar., 1892	36	—	172
,,	British Commissioner. Treaty. King Mwanga	29 May, 1893	App.	—	978
,,	Great Britain and Congo. Limits. Spheres of Influence	12 May, 1894	App. 16	—	1008

ALPHABETICAL INDEX.

Name of Country, Place, &c.	Subject.	Date of Treaty or other Document.	No. of Doc.	Art.	Page.
Uganda	British Protectorate	29 May, 1893	App. 12	—	995
,,	,, ,,	18 June, 1894	App. 17	—	1016
,,	Withdrawal of British East Africa Company	29 May, 1893	App. 12	1	995
,,	British Protectorate	27 Aug., 1894	App. 22	—	1023
Ukami	Treaty. German Colonization Society and Native Chiefs	1884	82	—	306
,,	Charter. Ditto	17 Feb., 1885	81	—	303
Umbe River	See Wanga River.				
Umtali	British Jurisdiction	27 June, 1891	App. 9	—	990
Umzimaritu	River. Pondoland. See Great Britain (Cape Colony).				
Umzimkulu	River. Pondoland. See Great Britain (Cape Colony).				
United States	Ratification. "Brussels Act"	2 Feb., 1892	22	—	102
,,	And Congo. Recognition of Association	22 April, 1884	64	—	244
Unyoro	Boundary. Uganda	18 June, 1894	App. 17	—	1016
Urivflan	Waterway. Rio del Rey Boundary	14 April, 1893	131	—	654
Usagara	Treaties. German Colonization Society and Native Chiefs	Nov.—Dec., 1884	81	—	303
,,	Charter. Ditto	17 Feb., 1885	81	—	303
Usoga	Boundary. Uganda	18 June, 1894	App. 17	—	1016
Usutu River	See Maputo.				
Uzeguha	Treaty. German Colonization Society and Native Chiefs	1884	82	—	306
,,	Charter. Ditto	17 Feb., 1885	81	—	303
Valvisch Bay	See Walfisch Bay.				
Victoria	See Ambas Bay.				
Victoria Nyanza	Lake. British and German Spheres of Influence	Oct.—Nov., 1886	123	3	617, 620
,,	,, ,, ,,	3—4 Dec., 1886	153	3	756, 759
,,	,, ,, ,,	1 July, 1890	129	1, § 1	642
Vintang	District. "British" Sovereignty and Protection	17 Sept., 1887	90	—	382

ALPHABETICAL INDEX.

Name of Country, Place, &c.	Subject.	Date of Treaty or other Document.	No. of Doc.	Art.	Page.
Vintang	Creek (Gambia). British and French Boundary..	10 Aug., 1889	110 Annex	1, § 1 2, § 5	558 565
Volta	British and German Limits. Interior ..	Dec., 1887	127	—	628
,, ..	Districts. British and German Boundary. (Map)	1 July, 1890	129	4	646
,, ..	River. British and French Boundary ..	26 June, 1891 12 July, 1893	114 118	— 4	574 591
,, .. ,, ..	" " " " East of Gold Coast and Togoland. Customs Union .. .. See also Great Britain (Gold Coast).	24 Feb., 1894	Note	—	661
Walfisch Bay..	British Occupation. Boundaries	12 Mar., 1878 14 Dec., 1878	89 89	— —	358 360
,, ,,	" " Annexation to Cape Colony	22 July, 1884 7 Aug., 1884	89 89	— —	360 360
,, ,,	" " " " Excluded from German Protectorate ..	5 Sept., 1884	83	—	360
,,	Southern Boundary. Arbitration	1 July, 1890	129	3	646
Wanga..	(Umba) to Rovuma. Coast between	28 Apr., 1888	198	—	238
,, ..	,, German East Africa Association. Zanzibar Dominions	14 June, 1890	155	8	763
,, ..	Zanzibar Possession .. See also Zanzibar Limits.	9 June, 1886	120	5	606
,, ..	To Kipini. Anglo-German Spheres ..	29 Oct.— 1 Nov., 1886	123	3, 4	616
,, ..	,, Concession. Zanzibar to British East Africa Company for 50 years	24 May, 1887	24	4	110
,, ,,	,, Royal Charter .. ,, Concession. Zanzibar to British East Africa Company for 50 years	3 Sept., 1888 9 Oct., 1888	25 26	— 1	118 126
,, ,,	,, Customs Revenues ,, Concession. Zanzibar to British East Africa Company "in perpetuity".. ..	21 Dec., 1889 5 Mar., 1891	29 31	— —	146 150
,, ..	,, Anglo-German Boundary	27 Oct.— 24 Dec., 1892	130	—	652

ALPHABETICAL INDEX.

Name of Country, Place, &c.	Subject.	Date of Treaty or other Document.	No. of Doc.	Art.	Page.
Wanga..	To Kipini. Anglo-German Boundary	25 July, 1893	132	—	656
,, ..	To Kismayu. British Administration of Coast	1 July, 1895	App. 36	—	1070
,, ..	To Congo State. British and German Spheres	1 July, 1890	129	1, §1	644
,, ..	River. Territory South of, excepted from British Protectorate See also Germany.	4 Nov., 1890	156	—	766
,, ..	To Victoria Nyanza. British and German Spheres	3, 4 Dec., 1886	153	3	756 759
,, ..		1 July, 1890	129	1	642
Wari ..	,, ,, ,, (Forcades.) See Great Britain (Niger).				
Warsheikh	See Benadir Ports.				
Waterways	Transit. Persons and Goods. British and Portuguese Spheres	11 June, 1891	150	12	737
West Africa	Settlements. See Great Britain (Sierra Leone)				
West Huk	Rio del Rey Boundary	14 April, 1893	131	2	654
Whemi..	(Lagos). Kingdom	23 July, 1886	82	—	425
Whydah	Limits. Sovereignty, &c.	—	65	Notes	248
,,	French Annexation See also Dahomey.	3 Dec., 1892	65	—	249
Witu ..	Sultanate. Independence from Zanzibar	June, 1885	82	—	307
,, ..	German Protectorate of Witu	2 Dec., 1885	82	—	314
,, ..	Cessions to German Subjects. Kipini to Witu	1885—1887	82	—	312 314
,, ..	Limit. Coast line North of Kipini to North of Manda Bay	Oct.—Nov., 1886	123	5	617 620
,, ..	Excepted from Lease to British East Africa Company of Zanzibar Dominions	31 Aug., 1889	154	1	760
,, ..	German Protectorate over Coast. Witu to Kismayu	22 Oct., 1889	32	—	315
,, ..	Right of Sultan to Manda and Patta Islands	20 Dec., 1889	82	—	316
,, ..	German Protectorate withdrawn	1 July, 1890	129	2	644
,, ..	British Recognition of Sultan's Sovereignty	1 July, 1890	129	2	645

ALPHABETICAL INDEX.

Name of Country, Place, &c.	Subject.	Date of Treaty or other Document.	No. of Doc.	Art.	Page.
Witu	Limits. Kipini to point opposite Kwyhoo ..	1 July, 1890	129	2	645
,, ...	German recognition of British Protectorate	1 July, 1890	129	11	650
,, ...	Coast between Witu and Kipini	19 Nov., 1890	82	—	313
,, ...	British Protectorate ..	19 Nov., 1890	82	—	316
,, ...	,, Witu up to Kismayu	19 Nov., 1890	86	—	327
,, ...	Submission of Witu to British Government	25 Jan., 1891	32	—	156
,, ..	,, ,, ,,	5 Mar., 1891	32	3	154
,, ..	Consent of British Government to Transfer to British East Africa Company ..	5 Mar., 1891	32	—	153
,, ..	Relations with Zanzibar ..	5 Mar., 1891	32	6	154, 155
,, ...	Administration of Witu by British East Africa Company ..	18 Mar., 1891	33	—	157
,, ...	,, ,, ,,	20 Mar., 1891	34	—	160
,, ...	British Protectorate over Provinces	20 Mar., 1891	34	—	163
,, ...	British Flag hoisted ..	April or May, 1892	Note	—	645
,, ...	Alcoholic Liquors prohibited. Witu to Kismayu	18 June, 1892	86	—	328
,, ...	Administration by British East Africa Company relinquished ..	31 July, 1893	160	—	770
,, ...	British Protectorate ..	31 July, 1893	160	—	770
,, ...	,, Delegated to Sultan of Zanzibar.	31 July, 1893	160	—	770
Wonkafong ..	Sierra Leone. Boundaries See also Great Britain (Sierra Leone).	29 Jan., 1852	100	—	503
Xesibe ..	Territories. Annexed to Cape Colony .. See also Great Britain (Cape Colony).	23 Aug., 1886	89	—	351
Ya Comba ..	(Sierra Leone). Cession to Great Britain ..	24 Sept., 1825	100	—	491
,,	Annexation to Sierra Leone	3 Oct., 1825	100	—	493
,,	British Sovereign Rights confirmed	18 Nov., 1882	100	—	520
Yarbatenda ..	(Gambia). British and French Boundary ..	10 Aug., 1889	110	1, § 1 § 2 Annex 2 § 3 & 6	558 559 564 565

ALPHABETICAL INDEX.

Name of Country, Place, &c.	Subject.	Date of Treaty or other Document.	No. of Doc.	Art.	Page.
Yelboyah	Island. British Possession	28 June, 1882	100	2	555
Yola	Interior. Gulf of Guinea. British and German Spheres	July—Aug., 1886	122	—	612
,,	,, ,, ,,	15 Nov., 1893	133	—	658
,,	Route. France and Germany. Limits	4 Feb., 1894	App. 13	—	998
Yorubaland	Non-cession of Territory, Boundaries, &c.	23 July, 1888	92	—	430
Zafarine	Islands. Spanish Possession	Jan., 1848	180	—	884
Zaire	River. See Congo River.				
Zambesi	Navigation. Great Britain and Portugal. (Not ratified)	26 Feb., 1884	147	—	713
,,	Germany and Portugal. Boundary. Catima Rapids	30 Dec., 1886	85	1	323
,,	German Access to, from South-West	1 July, 1890	129	3	646
,,	Free Navigation. (Not ratified)	20 Aug., 1890	148	—	722
,,	Free Navigation	14 Nov., 1890	149	—	728 729
,,	,, ,,	11 June, 1891	150	12, 13	737 738
,,	,, ,, Affluents	11 June, 1891	150	—	739
,,	North of. British and Portuguese Limits.. (Not ratified. See p. 726.)	20 Aug., 1890	148	1 5	715 718
,,	,, British and Portuguese Limits	11 June, 1891	150	5	734
,,	,, ,, ,,	May—June, 1893	151	—	743
,,	South of. British and Portuguese Spheres	20 Aug., 1890	148	2	716 718
,,	,, ,, ,,	11 June, 1891	150	2, 6	732 734
,,	Upper. British and Portuguese Spheres	11 June, 1891	150	3—6	733 734
,,	South of. British and Portuguese Mutual Right of Pre-emption	11 June, 1891	150	7	735
,,	Extension of Portuguese Sphere	20 Aug., 1890	148	3	717
,,	Territorial Limits. British and Portuguese	20 Aug., 1890	148	—	715
,,	,, ,, ,,	14 Nov., 1890	149	4	729

ALPHABETICAL INDEX.

Name of Country, Place, &c.	Subject.	Date of Treaty or other Document.	No. of Doc	Art.	Page.
Zambesi	Transit over Waterways and Landways	14 Nov., 1890	149	2	723
,,	,, ,, ,,	11 June, 1891	150	12	737
,,	Waters under Portuguese Influence	11 June, 1891	150	13	739
,,	,, under British Influence	11 June, 1891	150	13	740
,,	Watershed. Zambesi and Shiré. British and Portuguese Spheres	11 June, 1891	150	1	732
,,	Territory north of. See also British South Africa Company. France and Great Britain.				
Zanzibar	Limits. Possessions	3—4 Dec., 1886	153	—	754
,,	Brava. See Benadir Ports.				
,,	Kau. See Kau.				
,,	Kipini. See Kipini.				
,,	Kismayu. See Benadir Ports.				
,,	Lamu Island. See Lamu.				
,,	Mafia Island. See Mafia.				
,,	Magadisho. See Benadir Ports.				
,,	Meurka. See Benadir Ports.				
,,	Minengani River to Kipini. See Minengani River.				
,,	Pemba Island. See Pemba.				
,,	Tunghi Bay to Kipini. See Zanzibar Limits.				
,,	Warsheikh. See Benadir Ports.				
,,	Zanzibar Island. See Zanzibar Possessions.				
,,	Accession to "Berlin Act," with Reservations	8 Nov., 1886	193	—	925
,,	Reservations withdrawn	24 June, 1892	Note	—	925
,,	A Signatory Power to "Brussels Act"	2 July, 1890	18	—	48
,,	And Austria-Hungary. Consular Jurisdiction, &c.	11 Aug., 1887	194	—	926
,,	And Belgium. Consular Jurisdiction, &c.	30 May, 1885	195	—	926
,,	And British East Africa Company. See British East Africa Company and Zanzibar.				
,,	,, Zanzibar, Pemba, &c., excluded from concession to British East Africa Company	9 Oct., 1888	26	4	134

ALPHABETICAL INDEX.

Name of Country, Place, &c.	Subject.	Date of Treaty or other Document.	No. of Doc.	Art.	Page.
Zanzibar	And France. Consular Jurisdiction, &c.	17 Nov., 1844	196	—	927
,,	And Germany. Consular Jurisdiction, &c.	20 Dec., 1885	197	—	930
,,	And German East Africa Society. Concession. Mainland and South of the Wanga	28 Apr., 1888	198	—	933
,,	,, ,, Custom Houses, ditto	4 June, 1888	199	—	941
,,	,, Suppl. Agreement. Régie or Lease of Customs Duties to Society	13 Jan., 1890	200	—	943
,,	And Great Britain. Ex-territoriality, Consular Jurisdiction, &c.	30 Apr., 1886	152	16	751
,,	,, Treaties, &c.	1886—1893	152—160	—	770
,,	And Italy. Consular Jurisdiction, &c.	28 May, 1885	201	—	945
,,	,, Consular Jurisdiction subject to Italian Laws. (Additional Article)	10 Oct., 1885	201	—	947
,,	,, Transfer to Italy of Concession to British East Africa Co., Benadir Ports, &c.	8 Apr., 1890	202	—	949
,,	,, Concession, ditto	12 Aug., 1892	203	—	950
,,	,, Provisional Administration by Italy of ditto for three years	15 May, 1893	204	—	958
,,	And Muscat. Award. Independence of Sultans	2 Apr., 1861	205	—	961
,,	And Portugal. Consular Jurisdiction, &c.	25 Oct., 1879	206	—	963
,,	And United States. Ditto	21 Sept., 1833	207	—	965
,,	,, ,, ,,	3 July, 1886	208	—	966
,,	And Pemba. Limits. Maritime, Littoral, and Continental	9 June, 1886	120	—	605
,,	,, ,, ,,	Oct.—Nov., 1886	123	—	15—618
,,	,, ,, ,,	4 Dec., 1886	124	—	622
,,	,, ,, ,,	3—4 Dec., 1886	153	—	754—757
,,	International Bureau	2 July, 1890	28, 70—85	—	64, 78—82
,,	British Jurisdiction by other than Consular Officers	2 Feb., 1891	157	—	767

ALPHABETICAL INDEX.

Name of Country, Place, &c.	Subject.	Date of Treaty or other Document.	No. of Doc.	Art.	Page.
Zanzibar	British Jurisdiction by other than Consular Officers. Disputes. Zanzibar or other non-Christians and British Subjects	16 Dec., 1892	158	—	768
,,	,, Enforcement of Treaties in Zanzibar	17 July, 1893	159	7	769
,,	,, Natives of British Protectorates	17 July, 1893	159	—	769
,,	,, Over Zanzibar Subjects..	17 July, 1893	159	—	769
,,	Administration. British Protectorate north of the Tana	31 July, 1893	160	—	770
,,	Independence. Recognition by Great Britain and France	10 Mar., 1862	107	—	547
,,	,, ,, by Germany..	Oct.—Nov., 1886	123	7	318— 621
,,	,, ,, Agreement of 1862 modified	5 Aug., 1890	112	.—	570
,,	German Protectorate. Territories West of Zanzibar Possessions	6 Mar., 1885	82	—	805
,,	Treaties with Foreign Powers to be respected by British East Africa Company	24 May, 1887	24	14	110 114
,,	,, ,, ,,	9 Oct., 1888	26	4	130
,,	Relations with Foreign Powers	14 June, 1890	155	2	763
,,	Succession to Throne. British Guarantee..	14 June, 1890	155	45	764
,,	List of Sultans	—	—	Note	110
,,	German Protectorate over Coast up to Kismayu withdrawn	1 July, 1890	129	11	651
,,	Cession to Germany of German East Africa Company's Mainland Concessions and Island of Mafia	1 July, 1890	129	11	650
,,	Ditto. Recognition by Great Britain	27—28 Oct., 1890	129	—	650
,,	British Protectorate of Zanzibar Dominions (with exceptions)	14 June, 1890	155	1	763
,,	,, ,, ,,	4 Nov., 1890	156	—	766
,,	British Protectorate over Zanzibar, Pemba, and Witu recognised by Germany..	1 July, 1890	129	11	650
,,	Free Port	8 Feb., 1892	App. 10	—	992

ALPHABETICAL INDEX.

Name of Country, Place, &c.	Subject.	Date of Treaty or other Document.	No. of Doc.	Art.	Page.
Zanzibar	Import Duties. Free Trade. "BerlinAct"	22 June, 1892	App.	—	976
,,	Sultan's Dominions on Mainland to be Administered by British Officers	1 July, 1895	App. 36	—	1070
,,	Part belonging to Zanzibar to be under Sultan's Sovereignty, but under British Administration	1 July, 1895	App. 36	—	1070
,,	Mahommedan Law and Religion	1 July, 1895	App. 36	—	1070
,,	Religious Liberty..	1 July, 1895	App. 36	—	1070
,,	French recognition. German acquisition. Territorial possessions	1 July, 1895	App. 36	—	1070
Zeila	Non-conclusion of Treaties with Foreign Powers	3 Sept., 1840	178	—	832
,,	Customs Dues. Trade	7 Sept., 1877	67	2	261
,,	British and French Limits	2-9 Feb., 1888	App. 11	1	993
,,	British and Italian Trade, &c.	5 May, 1894	136*	3	670
Zones defined	Maritime. Congo Basins. "Berlin Act"	26 Feb., 1885	17	1, § 2	24
,,	Maritime. Repression of Slave Trade. "Brussels Act"	2 July, 1890	18	20 21	62 63
,,	Arms and Ammunition. Import prohibited	2 July, 1890	18	8	56
,,	Spirituous Liquors. Import prohibited	2 July, 1890	18	90	84
Zoolah	Cession to Great Britain	5 Oct., 1843	93	—	532
Zula	Italian Protectorate	2 Aug., 1888	11	—	10
Zululand	Boundary. Natal	5 Oct., 1843	92	—	434
,,	Boundaries. South Africa Republic	3 Aug., 1881	179	—	842
,,	" " "	27 Feb., 1884	179	—	849
,,	Boundary. New Republic	22 Oct., 1886	179	—	860
,,	A British Possession	14 May, 1887	102	—	533
,,	Notification to Powers	8 July, 1887	—	—	47
,,	Notes on	1843—1888	102	—	532
,,	Extension of Boundaries of Zululand	9 Dec., 1888	102	—	534
,,	Incorporation with, of part of Tongaland. See Tongaland.				
Zumbo	Settlement	20 Aug., 1890	148	—	716

www.ingramcontent.com/pod-product-compliance
Lightning Source LLC
Chambersburg PA
CBHW021814230426
43669CB00008B/752